Calcium

The importance of calcium in the diet has been highlighted
by the increasing incidence of osteoporosis amongst older
people. This book shows you how to take preventive
action by laying down strong bones in early life. It also
looks at calcium's other roles within the body.

Calcium

BEAT THE OSTEOPOROSIS EPIDEMIC

LEONARD MERVYN
B.Sc., Ph.D., C.Chem., F.R.S.C.

THORSONS PUBLISHING GROUP

First published 1988

©Leonard Mervyn 1988

British Library Cataloguing in Publication Data
Mervyn, Leonard, *1930–*
Calcium.
1. Man. Bones. Osteoporosis
I. Title
616.7'1

ISBN 0–7225–1579–0

Published by Thorsons Publishers Limited, Wellingborough, Northamptonshire, NN8 2RQ, England.

Printed and bound in Great Britain by
Collins, Glasgow

3 5 7 9 10 8 6 4

———Contents———

Dedication

To Charlotte Anne, on the threshold of life, with the hope that when her generation reaches mature adulthood, the conditions discussed in this book will have been relegated to the realms of medical curiosities.

─────Introduction─────

At the Inaugural Conference of the British Society for Nutritional Medicine held at St Bartholomew's Hospital in September 1984, Dr A St J Dixon, consultant physician at the Royal National Hospital for Rheumatic Disease, Bath, stated in his address that in his opinion, osteoporosis in Great Britain had now reached epidemic proportions. His statement simply echoed the growing concern, worldwide, that this insidious complaint was establishing itself as the cause of pain, bone fractures, and in some cases death, among the ageing population of the world — this problem afflicting mainly women but by no means restricted to them.

Osteoporosis is essentially honeycombing of the bone structure causing this to become weakened and prone to fracture in conditions where a healthy bone could withstand such trauma. The bone structure is composed mainly of calcium phosphate in one of its many forms, and the honeycombing effect of osteoporosis is due to loss of this mineral from the bone. Why calcium and phosphate are lost from the bone in osteoporosis is not entirely clear, but in any attempt to treat the complaint this loss must be replaced, reversed, or at least slowed down for any benefit to be gained. The role of calcium in these processes has been studied widely and whilst the effectiveness of calcium therapy alone is controversial there is no doubt that it must be supplied in the diet in adequate amounts no matter what other treatment is undertaken. The various aspects of calcium metabolism in osteoporosis treatment are dealt with in this book.

Prevention is better than cure and this axiom cannot be better illustrated than in the role of calcium deposition or mineralization within the skeleton in preventing the onset of osteoporosis. It is generally accepted that laying down strong bones in early life is a good insurance against the development of osteoporosis in later years so diet plays an important part in determining the extent of calcium deposition. At the same time there are indications that as women approach the menopause they change their diets because of an increasing tendency to put on weight and the first food items they cut down on, usually milk and milk-derived products, are in themselves the richest sources of the calcium they need. Thinner bones may therefore be exacerbated by a concomitant decrease in calcium intake. The dietary advice in this book will enable readers to become aware of foods other than dairy products that can supply their calcium needs so there is no reason why they should be deprived of the mineral.

Reducing the chances of developing osteoporosis can therefore be in part with the individual. The main determinant of bone density for many years after the menopause is the initial value at the time of menopause. This is because the range of bone densities in premenopausal women is large compared with the rates of bone loss. Hence the women who require preventive treatment are those whose bone density at the time of menopause is lower; they are at increased risk of fracture from the beginning and cannot afford to lose any more bone. (B E Nordin et al, 1987, *British Medical Journal*.)

Ensuring adequate intakes of calcium in the early part of life is therefore a measure that everyone can undertake. The second measure is to take adequate but not excessive exercise throughout life.

Calcium metabolism and osteoporosis are bound together irretrievably, but research carried out in only the last five years has suggested that calcium has other roles within the body. Preliminary studies indicate that adequate calcium intakes may help control blood pressure, keep blood cholesterol and fats figures down to normal levels, and perhaps be an important factor in preventing certain types of cancer. These exciting developments are discussed in this book and whilst the studies mentioned are in the early stages, there is no reason why increased calcium intakes cannot be taken now as potential prophylactic measures against these life-threatening conditions.

Minerals have long been considered the Cinderella of nutrition but with our increasing knowledge that they act in the body beyond their well-known traditional roles, their importance to our health continues to be underlined. It has always been assumed in the past that all our needs are met from modern diets. In the case of calcium at least, recent thinking is at last challenging this questionable concept.

CHAPTER 1

Osteoporosis——
——its causes and——
treatments

Before we consider the various aspects of osteoporosis, particularly in its relationship with calcium, it is important to understand how bone is formed in early life and the process involved in maintaining it once it is established.

Bone is composed of deposits of calcium phosphates within a soft, fibrous organic matrix known as collagen. Collagen is a protein and its fibres form cross-links that lay the foundation for the deposition and arrangement of hydroxyapatite crystals that are the major component of the inorganic portion of bone. It is the collagen fibres that give the bone support and tensile strength. Such properties make the bone resistant to breaking caused by simple falls or blows.

The main mineral, hydroxyapatite, is more commonly known as calcium phosphate and is available in supplements as bone phosphate or bonemeal. Hydroxyapatite also occurs in the earth as a natural mineral. In the early part of bone formation the calcium phosphate moiety is present more as an amorphous, non-crystalline type but as the individual grows this is superseded by a harder, more crystalline type of the mineral. Both types are present in the skeleton throughout life but the balance veers more towards the hard variety as age progresses. Calcium phosphate is also deposited in the teeth but its crystals are bigger than those in bone and so contribute to the greater hardness of tooth enamel and dentine.

Calcification

The process of bone building by deposition of these hard crystals of calcium phosphate is known as calcification or ossification. The process is of prime importance in the body of the growing child. The bones of an infant are like firm cartilage with a low content of calcium and phosphorus. They are hardened by the incorporation of the calcium-phosphorus (phosphate) complex. During this process, changes occur on both the inside and the outside of the bone because, as well as hardening, the bone must also grow. New bone tends to be formed around the shaft of a long bone. As this happens, bone material on the inside of the shaft becomes solubilized and transported elsewhere in a soluble state. By this two-way process the walls of the bone thicken, but at the same time the space inside that holds the bone marrow widens. Hence a structure is gradually built up that is strong, durable and yet light. The space inside the bones is not wasted as it is here that the red and white corpuscles of the blood are produced.

The most important cell in this process is called an osteoblast. This is a specialized bone cell that synthesizes the organic matrix or collagen, controls its mineralization with calcium phosphate, receives and translates information on mechanical forces undergone by the bone, and modulates the activity of other specific bone cells called osteoclasts.

Osteoclasts arise from the bone marrow and whilst their action is essentially equal and opposite to that of osteoblasts they depend upon them for many of their actions. After osteoblasts have performed their functions most are destroyed but those that survive change to cells called osteocytes that contribute to the integrity of the bone structure.

Throughout life, whatever its age, the skeleton is continuously being removed by osteoclasts — a process known as bone resorption — and replaced by osteoblasts. Osteoclasts resorb bone in microscopic cavities; osteoblasts then reform the bone surfaces, filling the cavities. This renewal of bone is most rapid in the young because these two activities are precisely linked to ensure bone mass remains constant. Remodelling or turnover of bone is continuous and under the regulation of mechanical and nervous forces, hormones, and local regulatory factors. Any upset in the delicate balance between bone formation and bone resorption can cause excess deposition of minerals when formation is dominant or honeycombing of the bone when resorption prevails. The latter condition is known as osteoporosis.

—— Cortical bone and trabecular bone ——

An important feature to bear in mind when considering how osteoporosis is caused is the fact that two major forms of bone exist. One is called cortical bone which is hard and compact and makes up the external envelopes of the skeleton. The second is known as trabecular or medullary bone which forms plates that cross the internal cavities of the skeleton. This is a soft, spongy type of bone. The proportions of cortical and trabecular bone vary at different sites of the skeleton. For example, the spinal vertebrae contain predominantly trabecular bone whilst the femur (thigh bone) contains predominantly cortical bone. Both types of bone respond differently to metabolic influences and their susceptibility to fracture differs markedly. This is why, as we shall see, some parts of the skeleton are more likely to fracture than others once osteoporosis sets in.

On the surface of bones the two types of cells function in sequence. Activation of osteoclasts is followed by about two weeks of resorption. There follows a reversal phase when for the next two or three months osteoblasts take over and make new bone. This complex process works in groups of cells called 'basic multinuclear units' in both cortical and trabecular bone. The continued activation of such units is necessary to regulate the body's calcium content, to repopulate the skeleton with new bone cells, or to repair minor structural damage.

Even after the bones have ceased to grow in length, they continue to thicken by radial growth until about the age of 30 years. There is a short transitional period of stability when the bone density stays constant, but eventually age-related bone loss begins. This is due to an uncoupling of the fine osteoblast-osteoclast balance and the bone-resorption cells, the osteoclasts, begin to dominate the bone-forming osteoblasts. Differences in the loss of cortical and trabecular bone also start to appear at this time.

During her expected lifespan, a woman will lose about 35 per cent of her cortical bone and 50 per cent of her trabecular bone. For men the figures are 24 and 34 per cent respectively. Cortical bone is the predominant type in the shafts of the long bones, whereas trabecular bone is concentrated in the vertebrae of the spinal column and also in the ends of the long bones. Since trabecular bone with its greater surface area is metabolically more active than cortical bone, it is more sensitive to upsets in mineral (ie. calcium) balance. Hence the bones where the trabecular form predominates are more likely to demonstrate loss of bone-mass and hence be more susceptible to fracture. Both types of bone lose mineral in a biphasic pattern. There is a protracted slow phase that occurs in both sexes followed by a transient accelerated phase that occurs in women after the menopause.

In the case of cortical bone, the slow phase of loss begins at about age 40 years in both sexes at an initial rate of about 0.3 to 0.5 per cent per year and increases with ageing until it slows or ceases later in life.

However, in postmenopausal women, an accelerated postmeno-pausal phase of cortical-bone loss is superimposed on this pattern to produce a rate of bone loss of 2 to 3 per cent per year. This high rate of loss occurs immediately after the menopause but it is not maintained and after 8 to 10 years, the original bone loss of 0.3 to 0.5 per cent per year is restored.

There are, however, marked differences in the patterns of loss in trabecular bone and cortical bone. In both sexes the onset of trabecular-bone loss occurs at least 10 years before that of cortical-bone loss. In the female, the extent of premenopausal trabecular-bone loss is much greater than the extent of cortical-bone loss. But although this rate of accelerated postmenopausal phase of trabecular-bone loss is greater in the initial phases, its duration is much shorter. These dif-ferences probably relate to the variable and gradual onset of female sex hormone deficiencies that are a feature of the immediate pre- and post-menopausal states.

What is osteoporosis?

Osteoporosis literally means 'porous bone'. Whilst the outer form of the bones does not change, the bones have reduced substance, becoming less dense. As the bone mass decreases bones become more susceptible to fracture. A fall, blow, or lifting action that would not normally bruise or strain the average person can easily break one or more bones in someone with severe osteoporosis. The spine, waist, and hip are the most common sites of osteoporosis-related fractures although as the disease is generalized it can affect any bone in the body.

When the bones of the spinal column (the vertebrae) are weakened by osteoporosis, a simple action like bending forward to make a bed or lifting something off the floor can be sufficient to cause a 'crush frac-ture' or 'spinal compression fracture'. These vertebral crush fractures contribute to loss of body height, causing pain and giving rise to the humped back known as 'Dowager's Hump'.

The occurrence of osteoporosis of the spine increases with age. A re-cent study of about 2,000 women showed X-Ray evidence of osteoporosis in the spine in about 29 per cent of those aged 45 to 54 years, 61 per cent of those aged 55 to 64 and a staggering 79 per cent of those aged 65 and older. Vertebral crush fractures are more common in women than in men and generally occur in women between 55 and 75 years of age. Similar figures are seen in the occurrence of wrist fractures due to osteoporosis. A simple fall when the individual reaches out to catch herself is usually the cause of a broken wrist (Colles fracture).

Osteoporosis is often the underlying condition causing a broken hip and as we see later it contributes to a considerable number of these — for example it is the direct cause in broken hips of more than 200,000 Americans over the age of 45 years each year. A fall from a standing position can fracture a hip weakened by osteoporosis. When the condition is severe even a change of posture or weight distribution is sufficient to fracture a hip which in turn leads to the fall. Those who have hip fractures due to osteoporosis are generally older than people who suffer spinal fractures. Women are far more likely to suffer the condition than men (see figures later) whether it occurs in the hip or in the spine, reflecting the denser bone mass usually prevalent in men.

——The extent of the osteoporosis problem——

The major consequence of having osteoporosis is that thin bones break easily either spontaneously or with minimal injury. We have seen that when bones fracture as a result of osteoporosis they do so most often at three sites: these are the wrist, the spine, and the hip. These three fractures have a combined incidence of 35 to 40 per cent in women over 65 years of age with grave clinical consequences.

Epidemiological studies carried out in Europe and Great Britain some 18 years ago showed that the fracture rate of the wrist and forearm in white women rises about 10-fold between 50 and 70 years of age reaching about 50 in every 10,000 women annually at the older age. Once over 60 years of age, 25 per cent of white women have evidence of vertebral crush fractures shown up by X-ray examination.

In terms of disease rate mortality and cost, however, the consequences of fracture of the proximal femur, ie the top of the thigh, are even more important. By the late 1970s, fracture of the neck of the femur (hip) ranked third in the list of use of non-psychiatric beds in England and Wales. Calculations from hospital in-patient inquiry rates show that 31,618 women aged over 45 years were admitted with this diagnosis in 1977. Deaths in these women at 16.8 per cent were about 20 times that expected for a population of this age.

When men of a similar age group were studied, admissions to hospital with hip fracture amounted to only 7,674 and these were all regarded as at the end stages of senile osteoporosis. The reason for this sex discrepancy in the age-specific fracture rates of the hip and top of the thigh is almost certainly postmenopausal osteoporosis. The total number of bed-days used by this disorder in both sexes was 677,597 in 1977 which represents occupation of 10 per cent of all orthopaedic and accident beds at a cost, by 1981 figures, of roughly £48 million.

For reasons not understood completely, the incidence of fracture of the hip joint has risen sharply in recent years even more than the overall increase of 48 per cent in bed use for this condition between 1968 and 1977. When the consequences of fractures of the wrist and spine are added to these, the total incidence is devastating for both the postmenopausal population and the resources of the National Health Service (*British Medical Journal*, 1982).

Similar figures, on a pro-rata basis, are found in the United States. One out of every three women and one out of every six men will have had a hip fracture by the time they reach extreme old age. This is the most serious type of fracture as it is fatal in 12 to 20 per cent of cases. Half of those that survive face long-term nursing care in hospital or home. The costs, both direct and indirect, of osteoporosis in the United States are estimated at 6.1 billion dollars annually.

In all countries, women have more fractures related to osteoporosis than men, and whites have more fractures than blacks. This is because of differing bone masses; these are 30 per cent higher in men than in women, and 10 per cent higher in blacks than in whites. The denser the bone mass as a woman reaches the menopause the less likelihood she has of developing osteoporosis. Reduction in bone mass is the most important reason for the increased frequency of bone fractures in postmenopausal women and in the elderly of both sexes. By the age of 70 years some 40 per cent of women will have had at least one postmenopausal fracture. Vertebral fractures occur most frequently in women aged 55 to 75 years with accelerated loss of trabecular bone. Hip fractures occur most frequently in older men and women who have slowly lost both cortical and trabecular bone mass.

————————Who is at greatest risk?————————

Let us now look at who is likely to be at the greatest risk of developing osteoporosis — but before we do it is useful to discuss the two types of osteoporosis hypothesized by two leading American doctors who are regarded as experts on the subject.

In an important paper published in the *American Journal of Medicine*, (1983) Drs B L Riggs and L J Melton postulated that there are at least two types of osteoporosis. Their concept is based upon clinical features, bone density and hormonal changes, and the relationship of the disease patterns to menopause and age.

Type I osteoporosis, also called postmenopausal, characteristically affects women within 15 to 20 years after menopause. Less commonly, men of the same age group acquire a form of osteoporosis virtually indistinguishable from the postmenopausal type. The main clinical manifestations are fracture of the vertebrae, the wrist, and the forearm, but increased tooth loss is also common. The vertebral or spine fractures are of the crush type associated with deformation and pain. The skeletal sites of all these manifestations contain large amounts of trabecular bone.

Those who suffer from Type I osteoporosis all demonstrate a rate of trabecular-bone loss some three times the normal with a rate of cortical-bone loss of nearly normal. During this accelerated phase of bone loss, the trabecular structure collapses, weakening the vertebrae and predisposing to easy fracture. By the time the individual seeks medical help, they have reached a 'burned-out' stage at which trabecular-bone loss has reached its maximum and further loss is minimal.

According to the hypothesis, Type I osteoporosis is caused by factors closely related to or exacerbated by the menopause. Such factors (discussed in detail later) include accelerated bone loss, itself dependent on hormonal changes and a reduced ability to activate vitamin D. These in turn cause decreased calcium absorption. This defect in calcium absorption may further aggravate bone loss. All women after the menopause are deficient in the sex hormones — the oestrogens — because the ovaries slow down their production. However, the blood serum levels of oestrogens in all postmenopausal women are similar, whether they have Type I osteoporosis or not. This means that other factors must interact with oestrogen deficiency to determine individual susceptibility to osteoporosis. What these factors are is still under study.

Type II or senile osteoporosis occurs in men and women of 70 years or older and is manifested mainly by hip and vertebral fractures although fractures of the upper arm, the skin, and pelvis are not uncommon. The vertebral fractures are not the crush type of Type I osteoporosis but are of the multiple 'wedge' type leading to Dowager's Hump (Dorsal kyphosis). 'Wedge' type fractures are caused by trabecular thinning associated with a slow phase of bone loss (unlike the accelerated phase of Type I osteoporosis) which leads to a gradual and usually painless vertebral deformation.

Bone-density values in the parts of the skeleton affected in Type II osteoporosis are usually in the lower part of the normal range and this suggests a loss of cortical as well as trabecular bone. These processes causing Type II osteoporosis may therefore affect virtually the entire population of ageing men and women. As the slow phase of bone loss progresses, an increasing number of these people will have bone-density values at levels likely to lead to fractures.

The two most important age-related factors in determining the development of Type II osteoporosis are decreased osteoblast function and impaired activation of vitamin D leading to decreased calcium absorption. However, the effects of all risk factors for bone loss encountered over a lifetime are cumulative. The residual effects of accelerated bone loss after menopause many years earlier (Type I osteoporosis) may explain why elderly women have a twofold higher incidence of hip fractures than elderly men even though the rates of slow bone loss are similar in both sexes. The switch in women from the fast menopausal loss to the age-related slow bone loss in the immediate postmenopausal woman accounts for the fact that there is no rise in the incidence of hip fracture at this time.

The major historical risk factors for osteoporosis are these:

Being a woman
When measured on the basis of incidence of vertebral fractures, women are likely to have six to eight times the number men have. The critical time, as we have seen, is postmenopausal ie within 20 years after the menopause. As well as a more rapid loss in bone substances, women start out with less bone density than men do. Bone density depends on the amount of bone made during growth and its subsequent rate of loss. It is becoming increasingly clear that insufficient accumulation of skeletal mass during young adulthood predisposes that person to fractures in later life as age-related bone loss develops.

Being white or Asian
White women are at higher risk of osteoporosis than black women and white men are at a higher risk than black men. South African blacks, despite a low calcium intake, rarely have severe osteoporosis and hip fractures in elderly blacks are only one tenth those in elderly whites. Some experts estimate that by the age of 65 years, a quarter of all white women have had one or more fractures related to osteoporosis. Similarly Asian women have the same risks as white. On the other hand, in New Zealand the arm bone mineral content of Polynesian women was found to be 20 per cent higher than for white women and the Polynesians had a much lower rate of hip fractures. The greater in-

itial bone density of blacks probably explains why they have fewer osteoporotic fractures than whites or Asians. There is also some evidence that blacks have a greater resistance to resorption of bone that allows them to accumulate more bone during growth and at the same time they tend to lose less bone during ageing.

Hereditary

Heredity plays an important part in determining an individual's propensity to osteoporosis. The development of strong bones with a high density can be a familial trait that is genetically determined. The role of heredity was also indicated in a study by Dr D M Smith and colleagues who measured bone density of the forearms of twins (reported in *Journal of Clinical Investigation* 1973). Identical twins (monozygotic or produced from one divided egg) had comparable bone densities but similar twins (dizygotic or produced from two separate eggs) varied significantly in their bone density, one from the other. There can thus be variation even within families.

Being elderly

As we have seen, age-related osteoporosis can affect the elderly of either sex. The risk of fracture increases as the degree of trauma or injury increases but this also happens as the ability of the bone to withstand trauma decreases. The main cause of age-related fractures is increased bone fragility because of bone substance loss. Fractures do not occur until bone density has fallen below the values found in young healthy adults.

The inclination of the elderly to fall is an independent risk factor for fractures, both because they tend to fall more often and also when they do they have increased chances of injury. In many, both factors are present. Serious injury often results because nerve-muscle coordination is impaired and slowed reflexes in the elderly reduce their ability to break the impact of the fall. In addition, falls are more frequent in the elderly because of failing vision, nerve diseases, arthritis of the lower limbs, and the use of sedatives and other drugs (*Age and Ageing* 1977).

Early menopause

This is defined as a condition in any female undergoing menopausal changes before the age of 45 years. It is regarded as one of the strong predictors for the development of osteoporosis. One of the commoner causes is removal of the ovaries in young women since studies have indicated that they have a lower bone density in later life compared with females of the same age with intact ovaries. This surgical menopause

accelerates bone loss to between 10 and 15 per cent in excess of that
expected for the bones of the limbs and 15 to 20 per cent of that normal for the vertebrae. In these women however, oestrogen replacement with oral preparations of the hormone prevents, or at least slows
down, these excessive losses of bone.

Being underweight

Although being overweight is not a good idea for many reasons of
health, it is an observed fact that obese people have less tendency to
osteoporosis than thin ones. This is seen in the extreme condition of
anorexia nervosa but only in those who do not exercise. Anorexic
women who were more physically active had denser bones than those
who were sedentary, according to a report from Dr N A Rigotti and colleagues in the *New England Journal of Medicine* (1984). Women suffering from anorexia nervosa show a reduction in the bone density of
the forearm, and vertebral fractures may also occur, according to the
same report. In such individuals, osteoporosis is probably related to the
lower oestrogen concentrations present, making the women comparable in hormonal terms to those with premature menopause. The
difference is of course that oestrogen levels in anorexics revert to normal once food is eaten and a normal diet maintained.

Why obesity protects against osteoporosis is not clear, but one
possibility is that the extra weight increases the stress of skeletal loading
so stimulating the bone to lay down more calcium. There is some
evidence that resorption of bone is decreased and oestrogen production is increased in those who are overweight.

Male menopause

Although men do not undergo the equivalent of the menopause in
defined clinical terms, it has been established that as they age, their
production of male sex hormones declines. Once this happens, the
chances of developing osteoporosis increase, according to a study
reported in *Hormone Research* (1984) by Dr C Foresta and colleagues.
In younger men with reduced testicular function a similar phenomenon is seen, suggesting that the male sex hormone testosterone has a
similar protective effect against osteoporosis as oestrogen has in
women. The reduced bone density measured in these men can be
restored to normal by the appropriate hormonal therapy.

Alcohol drinking

Excessive alcohol consumption can cause a reduced density of bones, according to a report by Dr H Spencer and colleagues reported in the *American Journal of Medicine* in 1986. Ethanol depresses bone formation by a direct effect on the osteoblasts. When 96 heavy drinkers attending a Veterans Administration Hospital in the United States were assessed in bone density, 45 had evidence of extensive bone loss. One third of these were between the ages of 31 and 45, suggesting that alcohol rather than age was the factor inducing the osteoporotic changes. Similar studies on 22 alcohol abusers indicated that six of them had evidence of osteoporosis with a reduction in the thickness of the trabecular-bone mass confirming that alcohol directly damages bone cells.

Cigarette smoking also appears to be a risk factor for osteoporosis although the evidence is not as well established as that for alcohol.

A chronically low calcium intake

One indication that long-term dietary calcium deficiency can reduce bone density has come from a study in Yugoslavia that was reported by Dr V Matkonic and colleagues in the *American Journal of Clinical Nutrition* in 1979. Residents of two Yugoslavian districts had an approximately twofold difference in dietary calcium intakes. Those residents of the district where calcium intake was lower had a reduced density in the bones of the hand compared to the other group and their incidence of hip fractures was higher. These differences were apparent in young adulthood but as their age progressed these criteria did not diverge between the two groups, suggesting that the effect of extra dietary calcium was on initial bone density. The study does underline the importance of adequate calcium intakes throughout life.

Medicinal drug therapy

Osteoporosis is common in people on long-term corticosteroid (cortisone) therapy. This is believed to be due to reduced activity by the osteoblasts coupled with stimulation of bone resorption as the result of a direct or indirect action by the drug.

Diuretic drugs of the thiazide type, on the other hand, are protective against bone loss. They are believed to act by increasing the re-absorption of calcium by the kidneys and so conserving the mineral.

Other drugs that have been observed to increase the chances of osteoporosis when used on long-term therapy include heparin (a blood anticoagulant) and anti-convulsant drugs (used in the treatment of epilepsy). The reasons are not known.

──────────────**Medical conditions**──────────────

Certain diseases can lead to bone loss. These include: Cushings syndrome due to excessive body levels of adrenal corticosteroids, either because of pituitary gland malfunction or therapy with high doses of synthetic corticosteroids, (see above); hyperparathyroidism (excessive hormone production by the parathyroid glands); certain forms of cancer (eg lymphoma, leukaemia and multiple myeloma); hypercalcaemia (excessive excretion of calcium in the urine); diseases of the liver, pancreas or intestine, leading to an inability to absorb calcium from the intestine; partial or complete gastrectomy; pregnancy.

──────────────**Other risk factors**──────────────

Include short stature and small bones; inactivity (discussed in detail below under Osteoporosis and exercise); nulliparity (never having given birth).

──────────────**Osteoporosis and exercise**──────────────

It is now generally accepted that moderate weight-bearing exercise is an important part of both prevention and therapeutic programmes for osteoporosis. It is now clear that inactivity leads to bone loss.

Many research studies have demonstrated that normal healthy people who are bedridden for long periods of time lose bone mineral rapidly. Studies have also shown that astronauts living in a state of weightlessness in space have a striking loss of bone mass. Whilst such inactivities may be due to the inability of muscles to work against each other and against gravity, the reasons for bone density decrease are found at the cellular level. It is here where there appears to be a decrease in osteoblastic activity plus an increase in osteoclastic activity (Drs R B Mazess and G D Whedon, 'Immobilisation and Bone', *Calcif, Tissue Int.* 1983). It is attractive to imagine the all-important osteoblast — which synthesizes bone matrix, mineralizes it and controls the activity of osteoclasts — is in fact sensitive to mechanical stimuli but this is by no means proven.

The best type and amount of principal activity to prevent osteoporosis has not yet been established but a modest programme of weight-bearing exercise is recommended for people of all ages. These include young women who are working towards reaching a high 'peak bone mass' in their mid-thirties — the critical period — as well as middle-aged and older women who want strong bones before the ravages of the postmenopausal period set in. Possibilities for weight-bearing exercises include: walking, hiking, racewalking, jogging, running, skipping, dancing of all types, gymnastics, tennis, squash, badminton, netball, basketball, volleyball, rowing, weight training, skiing, and cycling as long as some hard uphill work is involved. Swimming and yoga are good healthy pursuits but are not generally thought to be weight-bearing exercises.

Some exercise is good but more is not necessarily better when considering it as a weapon against osteoporosis. Dr B Krolner and colleagues in their classical studies reported in *Clinical Science* (1983) found that exercise could delay or reverse loss of trabecular bone from the vertebrae of postmenopausal women. This encouraging finding was tempered a little by a report a year later by Dr B L Drinkwater and her colleagues from the University of Washington, Seattle, USA, showing that some young female athletes had an abnormally low density of their vertebral bones. However, not surprisingly, the frequency, duration, and intensity of training regimes for endurance athletes far exceed the physical activities of postmenopausal women, so perhaps the different observations in the two studies can be attributed to these variations. Also to consider is the oestrogen status of the two groups of women since, as we shall see, this has a bearing on the extent of bone loss in the younger age group.

Amenorrhoea

There have been a number of studies comparing the bone mineral contents of amenorrhoeic (non-menstruating) and eumenorrhoeic (menstruating) female athletes and it is relevant at this point to discuss two of them.

Amenorrhoea is the absence of menstrual cycles, characterized by low-circulating levels of oestrogens similar in some ways to the hormonal changes of the menopause that lead to accelerated bone loss. It is of concern that the condition of amenorrhoea in young athletes could place them at higher risk of osteoporosis in later life. The condition is widespread among athletes and ballet dancers. For example, in highly trained endurance athletes 25 to 40 per cent of the women report less than three menstrual periods per year.

Amenorrhoea has long been considered a rather benign side effect of endurance training in female athletes, but concern about the clinical implications of the phenomenon centred originally on possible detrimental effects on reproductive function. Now attention has switched to a potential adverse effect on bone mass. When we look at the incidence of amenorrhoea we find it affects 50 per cent of competitive runners; 44 per cent of ballet dancers; 25 per cent of non-competitive runners but only 12 per cent of swimmers and cyclists. The reason for the increased prevalence of amenorrhoea with weight-bearing exercise is not known.

The first study was carried out at the University of Washington and the Veterans Administration Medical Center, both in Seattle, USA, and reported in the *New England Journal of Medicine* (1984). Twenty-eight women athletes participated in the study; fourteen of them were amenorrhoeic, having had no more than one menstrual period in the preceding 12 months. The fourteen eumenorrhoeic women were chosen from a larger pool of athletes to match amenorrhoeic subjects in the following variables and in order of priority: sport, age, weight, height, and the frequency and duration of daily training sessions. Eleven of the subjects in each group were runners, the remaining three were crew members. All the women were non-smokers, in good health, and had not taken the contraceptive pill during the preceding six months.

On four separate occasions at seven-day intervals, each subject donated a sample of blood for measurement of female sex hormones (ie oestrogen, oestradiol and progesterone) and male hormone assay (testosterone). Bone density measurements were made of the forearm, wrist, and lumbar vertebrae. Each subject was accurately weighed and estimations made, using standard formulae, of total body fat.

The training regimes and physical characteristics of the two groups of athletes were similar but there were differences in the number of miles run per week. The amenorrhoeic participants ran an average of 42 miles, their eumenorrhoeic counterparts averaged 25 miles per week. Despite this difference, both types of athletes had a similar percentage of body fat. There were, however, differences in bone and sex hormone levels between the two groups.

Neither bone mineral content nor bone mineral density at the two sites investigated on the forearm showed any variability between the two groups of women. However, the minimal density of the lumbar vertebrae was significantly lower in the amenorrhoeic group of athletes. The bone mineral content and bone mineral density were found to be related in all cases studied but there was no significant relationship between the bone mineral density of the vertebrae and that at the forearm sites.

Not surprisingly, oestrogen levels in the non-menstruating women were significantly lower than in the others. Testosterone levels in both groups were similar to those expected in women of their age range.

The dietary intakes of the two groups were similar but there were some differences in total energy and fat eaten. The amenorrhoeic group had low intakes on both counts but statistically there was no significance in the figures. All the athletes studied had dietary calcium intakes exceeding the US RDA of 800mg daily for women of this age group.

What emerged from this study was that neither physical activity nor menstrual state resulted in any marked deviations from the normal mineralization of the predominantly cortical bone of the forearm. The difference lay, however, in the bone density of the predominantly trabecular bone of the vertebrae. Whereas the density in the women menstruating normally was close to that predicted for their age group, the average bone density of the amenorrhoeic athletes was equivalent to that of women of 51.2 years of age. Two of these athletes had a bone density lower than that where fractures would be expected to occur easily.

How do these findings relate to calcium balance, if indeed they do? It is generally accepted that low oestrogen levels after the menopause or in pre-menopausal women (as athletes are) with hormone dysfunction are related to the loss of bone mass in these groups. What is not certain is the role of oestrogen in the build-up and breakdown of bone itself. There are no oestrogen receptors in bone so any effects of oestrogen are assumed to be indirect. One such indirect route may be the effect of oestrogen on calcium balance.

There is ample evidence, exemplified by a paper in *Clinical Investigation Medical* (1982) by Dr R P Heaney, that lack of oestrogen increases the daily calcium requirements of women. Both sets of female athletes in the above *New England Journal of Medicine* study had the same dietary intake of calcium of at least 800mg per day. However, in the Heaney study it was found that in oestrogen-deficient women calcium absorption is decreased and its excretion is increased. Consequently he suggested a daily intake of 1,500mg calcium to maintain calcium balance in women with low oestrogen levels. Applying this suggested figure to the women athletes in the trial, it became obvious that the amenorrhoeic counterparts were not getting this. Differences in calcium balance between the two groups was as much as 30mg per day.

In conclusion then, it has been demonstrated that the amount of physical activity undertaken by the non-menstruating athletes in the study did not protect them against loss of vertebral bone. Whilst this does not negate the value of exercise in maintaining a healthy skeleton there must be some interaction between oestrogen and exercise in their effect upon specific skeletal areas which in turn reflects the type of bone present. In women, cortical bone (as in the forearm) is unaffected; trabecular bone (as in the vertebrae) is decreased.

————Male athletes — a different story————

Strangely, a different picture emerged from studies on male endurance runners. When Drs N Dalen and K E Ollson measured bone density in the cortical bone structures of male cross-country runners, they reported in the *Medical Journal Acta Orthop Scand* (1974) that the runners had a 20 per cent higher mineral content than non-athletic men of the same age group. There were no differences in the bone density of the lumbar vertebrae. In another study on male marathon runners (Dr J F Aloia and colleagues, *Metabolism*, 1978), these adults had higher cortical bone mass than sedentary men of the same age group. There is no hypothesis at present to explain these differences in the effect of excessive exercise on male and female skeletons. What is certain is that the female sex hormones play their part in some way unknown. However, whilst the questions are being answered regarding the long-term effect of amenorrhoea on the skeletal integrity of female athletes, it would appear to be a wise precaution for these women to ensure more than adequate calcium intakes. Heaney recommends 1,500mg daily, which is theoretically possible but difficult to achieve in practice from food alone and supplementation with the mineral is virtually mandatory. At least such an intake should help the woman to maintain positive calcium balance.

A similar study was reported in the *American Journal of Clinical Nutrition* (June 1986) by Miriam E Nelson and her colleagues from the USDA Human Nutrition Research Center on Ageing, Tufts University, Boston, USA. Twenty-eight women runners were involved of whom eleven were amenorrhoeic; the remainder were eumenorrhoeic. Bone mineral densities, oestrogen blood levels, maximal aerobic power, body-density measurements, and dietary intakes were all assayed as in the Seattle trial above. All diets were analysed for energy, protein, fat, carbohydrates, calcium, phosphorus, and dietary fibre.

Both groups of athletes had similar weight, height, and weight-height ratio and were within their desirable weight-for-height range. The calculated percentage of body fat was similar for both groups but there was a negative correlation between bone-mineral density of the lumbar vertebrae and percentage of body fat. This means that the body density in these subjects was affected by changes in the skeleton density as well as by changes in fat content. Usually in middle-aged women a large loss of bone substance (30 per cent) increases the apparent body fat by 9 per cent. In this study, though, differences in fat content were so minor that any decrease in body density could only come from bone loss. As before, there was no difference between the two groups in bone-mineral density of the forearm nor was this any different from those reported in sedentary women of the same age group. This is not unexpected as the forearm is not stressed during running and it has a high proportion of cortical bone. Moreover these findings confirmed a previous study (K P Jones et al, *Obstet, Gynaecol* 1985) in which amenorrhoeic athletes had a significantly higher bone-mineral content of the forearm when compared to amenorrhoeic non-athletes. The Jones finding is an important one since it indicates that physical exercise in amenorrhoeic women may have a protective effect on the bone-mineral content of cortical bone like that present in the forearm.

The Tufts University study also confirmed that the lumbar vertebrae of amenorrhoeic athletes had a significantly lower bone mineral density. When compared to sedentary women of the same age group, amenorrhoeic athletes consistently demonstrated lower bone-mineral values of the vertebrae whereas eumenorrhoeic athletes had comparable values to the non-athletes. Although the lumbar spine is stressed during weight-bearing exercise and hence would be expected to show an anabolic response to training, there was in fact no difference between the two groups. Any reduction in bone-mineral density must therefore be due to the mainly trabecular bone of the vertebrae being more susceptible to hormonal change.

The average concentration of circulating oestradiol (the most important oestrogen) in the amenorrhoeic women is only one-third that of eumenorrhoeic women in the early part of the menstrual cycle. This difference in oestrogens could significantly affect bone metabolism by changing calcium metabolism or by interfering with osteoblast activity. In addition, when oestrogen levels over long periods are compared, the amenorrhoeic athletes demonstrate low circulatory levels for a prolonged time. These in turn could be preventing a large peak-bone growth leading to a premature loss of bone mineral.

No differences were found in physical fitness, miles of runner per week, training pace, or years of training for both groups of women. This contrasts with the *New England Journal of Medicine* study in which the amenorrhoeic women were found to train more intensely than their eumenorrhoeic counterparts.

It was, however, in dietary intakes that significant differences between the two groups were found. The reported daily intake of energy was subnormal in the amenorrhoeic group. The percentage of energy supplied by fats, protein, and carbohydrates was identical in the two groups but the total energy intake was significantly lower in those athletes who were amenorrhoeic. In its make-up, their diet was similar to that of eumenorrhoeic athletes and sedentary women of the same age group but the amounts eaten were inadequate — 25 per cent less calories. There was no significant difference in energy expenditure during training so the amenorrhoeic women would have a 25 per cent lower energy balance. This compared well with the Seattle study although these researchers reported no differences in energy intakes between their two groups of women. Lower energy balance in their case could only have come from the more intensive training reported in the eumenorrhoeic group.

The results would suggest that amenorrhoeic athletes have an eating-behaviour problem, a phenomenon that is common in ballet dancers who also show a high prevalence of low food intake, delayed onset of menstruation and menstrual irregularities. It is therefore possible that a low intake of food maybe causally related to amenorrhoea. Hence amenorrhoea may be a functional adaptation to a negative energy balance whether it is produced by low energy intake alone or by increased energy expenditure together with a low energy intake as in the Tufts University study.

The average calcium intake in both groups involved in this study was higher than the US RDA of 800mg daily but individually there were very wide variations. There was however no correlation found between dietary calcium intake and bone-mineral content found in the amenorrhoeic group, so it could not be ascribed to a low calcium content in their diet. Much more likely is the marked blood oestrogen level differences between the two groups.

Low oestrogen levels in themselves lead to increased urinary calcium loss, at least in middle-aged women and during the menopause, according to Dr R P Heaney and colleagues (*J Lab Clin Med*. 1978 and *J Clin Nutr* 1977). In addition lower intestinal absorption of the mineral and an increased daily calcium requirement are also features of this age group. It is reasonable to imagine, therefore, a similar state of affairs in younger women who are amenorrhoeic because of their athletic pursuits. As the Tufts University study showed, their carbohydrate level intake was also low, another factor that could lead to reduced calcium absorption. On the positive side, though, their low protein intake would be expected to protect against excessive urinary calcium loss (Dr H Spencer et al, *American Journal Clinical Nutrition* 1978). Fibre and phosphate intakes in both groups of athletes were similar so these two food factors apparently did not contribute to reduce calcium absorption.

The conclusions of the Tufts study suggest that amenorrhoea in athletes may have deleterious long-term health consequences by reducing the mineral content of trabecular bone many years before the onset of the menopause and old age. In addition, the data suggests that an abnormally low food intake contributes to the disorder. It is evident that amenorrhoea in women athletes is not a problem solely for the gynaecologist but should also be evaluated and treated as a dietary disorder.

What conclusions then can we reach from these and other studies (eg C E Cann et al, 1984 *Journal of the American Medical Association*)? There is little doubt that some weight-bearing exercise can have a beneficial effect in reducing the chances of osteoporosis in all women but particularly in those approaching or going through the menopause. In one survey, women aged 36 to 65 years who took a 50 minute aerobics class three times a week lost only 2.5 per cent of the density in their forearm bones compared with 9.5 per cent for women who did not exercise. On the other hand, exercise that is strenuous enough to induce amenorrhoea in younger women can cause mineral bone-loss creating a premature kind of osteoporosis. If a serious woman athlete is menstruating normally then she has little to worry about regarding her bone-mineral density.

What can these amenorrhoeic women do to overcome their increased chances of low bone-mineral density? One approach is to eat more calories but keep the proportions of fat, carbohydrate, and protein the same to bring them up to the calorie intake of their eumenorrhoeic colleagues. Extra calcium to bring them up to 1,500mg daily intake may help (R P Heaney, *Clinical Journal Medicine* 1982) since in all the studies so far, although average calcium intake of the groups was above the US RDA of 800mg, many of the women were below this. It is difficult to achieve 1,500mg calcium from diet alone so supplementation becomes mandatory. Hormone replacement therapy may be prescribed in some cases of young female amenorrhoeic athletes and it is possible that taking the contraceptive pill may help since the synthetic oestrogens in these preparations are likely to have a protective effect against the development of reduced bone-mineral density (Drs N F Goldsmith & J O Johnston, 1975, *Journal of Bone Joint Surgery*).

————Calcium therapy in osteoporosis————

The fundamental feature of osteoporosis is loss of bone mineral and as this is essentially calcium phosphate it would appear that logically, the therapeutic approach would be to replace that lost calcium. Early studies confirmed this. For example, Dr R R Recher and colleagues (*Ann Internal Medicine* 1977) and Dr A Horsman and associates (*British Medical Journal* 1977) measured the bone density of the hand and forearm in postmenopausal women before and after calcium supplementation. They reported that a daily intake of between 1.5 and 2 grams of calcium reduced bone loss in the bones of these women. At the same time, in the Recher study, calcium balance investigation revealed that the bone resorption rate was also reduced. In order to maintain calcium balance after the menopause a daily supplement of 1 gram of calcium in addition to normal dietary intakes is required. This amount of extra calcium was claimed to produce 40 per cent fewer fractures in postmenopausal women compared with those who were untreated (Dr B C Riggs et al, 1982, *New England Journal of Medicine*). It must be pointed out, however, that the most effective treatment in this trial turned out to be a combination of calcium, fluoride, and oestrogens although each therapeutic agent alone reduced fracture rates to some extent. Vitamin D was ineffective.

A series of studies by Dr R P Heaney reported in 1976 (*Clin Pharmacol Ther*), 1977 (*American Journal of Clinical Nutrition*) and 1978 (*J Lab Clin Med*) looked at the possibility that calcium supplements alone might restore a long-standing deficiency. The papers quoted reported studies on 168 normal, well-fed women at around the menopause during five-year intervals. Their average intake of dietary calcium was 600mg daily, well below the recommended daily intake of 800mg in the United States. The group as a whole was in negative calcium balance — ie more calcium was being lost than absorbed. It was discovered that the higher the calcium intake the less negative the calcium balance. In fact, in order to maintain calcium equilibrium a total calcium intake of at least 1 gram a day before the menopause and

1.5 gram a day after it is needed. These quantities of the mineral ensured calcium losses did not exceed uptakes and by inference this positive calcium balance must lead to increased bone calcium.

The classic calcium balance studies reported above were extended more recently by Dr R P Heaney and his colleagues (*Amer J Clin Nutr* 1982) and they produced considerable evidence that calcium is good for the young premenopausal skeleton. Low calcium intakes, producing persistently negative calcium balances, contributed to bone loss. Results such as these were assessed by members of the Consensus Development Panel on Osteoporosis, organized by the National Institutes of Health of the USA and reported in the *Journal of the American Medical Association* in August 1984. They concluded that calcium metabolic studies indicate a daily requirement of about 1,000mg of calcium for premenopausal women and those undergoing hormone replacement therapy. Postmenopausal women not receiving hormones require about 1,500mg daily for calcium balance. The panel also decided that there were studies showing that high dietary calcium intake suppresses age-related bone loss and reduces the fracture rate in patients with osteoporosis. It suggested a calcium intake of 1,000 to 1,500mg daily beginning well before the menopause to reduce the incidence of osteoporosis in postmenopausal women. Similar recommendations for men may prevent age-related bone loss in them also.

Calcium therapy is only one of the regimes aimed at stimulating bone formation and increasing bone density substantially. Sodium fluoride is another mineral that may help, and it was in 1961 that it was first used in treating osteoporosis (Dr C Rich and J Ensinck, *Nature*). It improved the mineralization of bone but did not reduce the number of bone fractures. Reduction of these was achieved however once it was realized that fluoride is more effective when given with calcium.

Fluoride

Fluoride is a trace mineral and like many trace minerals a little is essential but too much can be harmful. A similar situation exists in the teeth where fluoride at a level in water of 1 part per million can protect against dental caries but too much causes fluorosis, a condition that is very damaging to teeth. When fluoride is taken at the intake of 30mg daily along with 1,000mg of calcium it was claimed to be protective against osteoporosis according to T Hansson and B Ross, 1982, quoted in *British Medical Journal* of 26 March 1983, p 100.

Fluoride has been shown to stimulate osteoblasts directly. In the doses needed it can increase the density of trabecular bone in the spinal column but cortical bone in the forearm shows no increase in mineral density even when it is given at the same time as calcium. This newly-formed fluoritic bone is less strong than the equivalent amount of normal bone but the substantial increase of trabecular bone-mass after fluoride plus calcium therapy does increase bone strength. The occurrence of new fractures after this treatment is reduced by a significant 64 per cent.

Fluoride-calcium combined therapy does not help in all cases of osteoporosis. Only one-half of patients have substantial increases in bone mass; one quarter have a partial response, but a quarter do not respond at all. Sodium fluoride is a toxic substance that should be taken only under medical supervision. Thirty per cent of those treated have symptoms of gastric irritation, with a further 10 per cent suffering acute pain in the lower abdomen. Fluoride is not as yet approved by any licensing authority as a treatment for osteoporosis as it is still being assessed for effectiveness and safety. If proved in two ongoing trials, it may soon be on the approved list, at least in the United States.

The hope that high calcium intakes may be enough in themselves to prevent and treat postmenopausal osteoporosis has not been confirmed by recent studies from Denmark. In the first of these, reported in the *British Medical Journal* of the 27 October 1984 by Drs L Nilas, C Christiansen and P Rodbro, a series of 103 early postmenopausal women completed a questionnaire about their dietary calcium intake. On the basis of this they were divided into three groups: those with an intake below 550mg per day, those with an intake between 500 and 1,150mg per day and those with an intake above 1,150mg daily. Each woman was then given a daily supplement of 500mg calcium for two years and had the bone mineral content of their forearms measured every three months. All the women had had a normal menopause six months to three years before and had not been taking any drug therapy known to influence calcium metabolism.

Over the period of two years' treatment all three groups showed a similar fall in bone mineral content despite total calcium intakes varying from 1,000mg to 2,000mg daily. Fourteen of the participants showed a positive change in bone mass but all of the others lost bone mass consistently. There was no correlation with calcium intake since the fourteen with higher bone mass took 1,390mg of the mineral daily that was not significantly different from the 1,520mg taken by the fourteen with the greatest bone loss. The individuals with the highest dietary intake of calcium (2,340mg daily) had a bone mineral loss rate of 4.5 per cent per year despite all that calcium.

One possible objection to the results reported in this trial is that the amount of supplementary calcium (500mg) is insufficient to be an effective dose in preventing bone loss in early menopause. A follow-up study was therefore carried out by Drs B Riss, K Thomsen and C Christiansen at the Department of Clinical Chemistry, Glostrup Hospital, Denmark, to compare the effectiveness of 2,000mg oral calcium daily with hormone replacement therapy in preventing bone mineral loss after the menopause. As in the first trial, all participants had their forearm and vertebral bone densities measured every three months, over a treatment period of two years. The patients were treated with calcium or oestradiol hormone or placebo (a harmless, non-therapeutic treatment). The trial was double-blind, which means that neither patients nor doctors knew which treatment was given to whom until the trial was finished. The results confirmed the previous trial in that treatment with calcium did not significantly reduce bone loss any more than the placebo treatment did. There was however an observed tendency towards a slower loss of cortical bone (forearm) than in the placebo group, but the loss of trabecular bone in the spine was the same as that in the group receiving placebo. Bone density remained constant in the oestradiol treated group so here at least further losses of bone mineral were halted.

What, then, can we conclude about the effectiveness of calcium supplementation in preventing osteoporosis in postmenopausal women? The evidence suggests that calcium alone is not as effective as oestrogens alone in preventing postmenopausal bone loss. However, treatment with 1,000mg calcium daily does appear to reduce bone resorption in the short term. (Dr M Horowitz et al, *Ann J Clin Nutr* 1984) and this reduction can persist for at least two years (Dr R R Recker et al *Ann Intern Med* 1977). We have seen above that calcium supplementation can reduce to some extent cortical bone loss although trabecular bone is unaffected. Cortical bone loss is a predictor of the risk of hip fracture so a reduced cortical bone loss may protect against such fractures. There are studies indicating that calcium supplementation protects against even spinal fractures (eg Dr B E C Noradin et al, 1980, *British Medical Journal*). Hence, extra calcium should be part of everyone's defence against osteoporosis as they approach middle age. What has not been established is that additional calcium can reverse an existing osteoporosis — in fact modern research suggests that no existing therapy can do this. Even hormone replacement therapy, preferably with extra calcium, will only halt bone loss and prevent exacerbation of the osteoporotic degeneration. The latest word on the role of calcium in preventing bone mineral loss has come from the National Institutes of Consensus Development Panel on Osteoporosis. In the 1984 meeting, discussed above, they concluded that everyone should consume 1,000mg calcium per day and that postmenopausal women should consume 1,500mg. At another meeting on osteoporosis held in January 1987 there was speculation that researchers would pull back from their previous recommendations. It did not happen despite the intervening reports on the lack of effectiveness of calcium alone in preventing mineral bone loss.

The conference 'reiterated that all adults should consume 1,000mg per day as part of a programme to prevent debilitating bone fractures in old age'. They also stood by the recommendation that postmenopausal women get 1,500mg of calcium per day. They also stressed the importance of oestrogen (hormone replacement therapy) for women at high risk, beginning soon after menopause. This approach, like that of using fluoride and the hormone calcitonin is purely the province of the medical practitioner so is unsuitable for self-treatment. However individuals can help themselves by ensuring they take the calcium intake recommended, preferably with vitamin D to ensure absorption of the mineral, and by partaking of weight-bearing exercise on a regular basis.

Calcium therapy in steroid-induced osteoporosis

We have seen that ageing and the menopause whilst the most common factors in the development of osteoporosis are not the only ones, and the long-term use of steroid (ie corticosteroid or glucocorticoid) drugs can lead to the condition. Steroid drugs are widely used in the treatment of conditions such as asthma and arthritis so they represent an important risk factor for developing osteoporosis in those taking them. Steroids inhibit bone formation and stimulate bone resorption by both direct and indirect actions.

Treatment with vitamin D alone has been claimed to restore calcium absorption to normal with a consequent decrease in bone resorption and increase in bone mass according to Drs T J Hahn and B J Hahn reporting in *Seminar Arthritis Rheum* (1976). As calcium absorption in general can also be increased by extra dietary intakes of the mineral, it might be expected that such intakes can produce similar beneficial effects upon calcium metabolism and lead to a decrease in bone resorption in those with steroid-induced osteoporosis. This theory was put to the test by Drs I R Reid and H K Ibbetson of the Department of Medicine, Section of Endocrinology, University of Auckland, Auckland, New Zealand, and reported in *The American Journal of Clinical Nutrition*, August 1986.

Thirteen patients (six men, seven women) who were receiving chronic prednisone therapy for various conditions (average 15mg daily for 3 years) were given 1,000mg calcium supplement daily which they took at night. Bone density measurements were not performed but calcium excretion in the urine and other tests were carried out to determine rate of bone mineral loss. Treatment with calcium and testing of the urine were assessed over a period of 2 months.

The results demonstrated that calcium supplementation reduced bone resorption, based on an increase in calcium excretion (proving improved absorption) and various blood and urine analyses, indicating that in addition to this reduced bone resorption there was no decline in bone formation. All of these factors would suggest an increase in bone mass produced by the extra calcium eaten during the test period. More studies are required, including direct measurements of bone density, but meanwhile it is a sensible approach for anyone on long-term administration of steroids to take extra calcium at a level of 1,000mg elemental calcium daily to reduce their chances of developing osteoporosis induced by the drug they are taking.

———Hormone therapy in osteoporosis———

There is no doubt that deficiency of circulating oestrogens is a major factor in the development of postmenopausal osteoporosis. In retrospective studies, Dr T A Hutchinson and colleagues showed that postmenopausal oestrogen therapy protects against fractures of the hip and forearm in osteoporotic women (*Lancet*, 1979). Confirmation came from studies reported in the *New England Journal of Medicine*, 1980, by Dr N S Weiss et al and in *Annals of Internal Medicine*, 1985, by Dr B Ettinger et al. Evidence has also been presented by Ettinger that postmenopausal bone loss is reduced by giving oestrogen therapy. A protective role for oestrogen against bone loss is also suggested by the finding that frequent pregnancies and subsequent lactation reduce bone loss and the risk of osteoporotic fractures (Dr J F Aloia et al, 1985, *American Journal of Medicine*), since oestrogen levels are high at these times.

The optimal duration of oestrogen treatment remains to be determined but oestrogens not only conserve bone mass when first started some years after the menopause but are effective in the long term also. Even after withdrawal of treatment the loss of bone in osteoporosis is not accelerated, which means that short periods of therapy given immediately after the menopause may give lasting protection to the skeleton. The value of long-term oestrogen replacement has been verified in young women who have had their ovaries removed surgically. Without hormones they have a lower bone density in later life than their peers with intact ovaries; with oestrogen therapy loss of both cortical and trabecular bone is prevented or slowed down. (Dr R Lindsey et al, 1980, *Lancet*).

How safe is oestrogen replacement therapy? It is well known that there is an increased incidence of cancer of the womb to about 1 per cent per year in those taking oestrogens after the menopause but this is usually manifested at an early stage and can be managed adequately in most cases (1984 National Institutes of Health Consensus Development Conference on Osteoporosis, *Journal of the American Medical Association*). There is no increased incidence in those who have undergone hysterectomy (removal of the womb).

It has also been claimed that prolonged oestrogen treatment can cause an increase in breast cancer, thrombosis in the veins, blood clotting in the lungs, high blood pressure, heart attacks, strokes, and gall stones. Apart from gall stones, however, it is probable that the risks of these occurring have been overestimated since postmenopausal women taking oestrogens appear to have a lower overall mortality than those not receiving treatment, according to Dr B E Hillover and colleagues reporting in the *American Journal of Medicine*, 1986. Once these findings are confirmed the risks of oestrogen therapy may well prove unfounded.

Some progress has already been made. Recently a promising study reported in the *New England Journal of Medicine* (9 October 1986) by Dr M L Padwick and his colleagues from Kings College School of Medicine and Dentistry, London, and the Chelsea Hospital for Women, London, confirmed that the risks of developing thickening of the womb and cancer of the same organ can be reduced by giving progesterone-like hormones along with the oestrogen. The combined therapy uses hormones similar to the active constituents of the contraceptive pill but at lower potencies. In addition, two controlled studies to evaluate the value of the combined therapy both indicated that it is effective in increasing bone mass when started within three years of the menopause. This beneficial response was maintained in other studies for the duration of three and ten years' treatment.

To conclude, it appears that low dose oestrogen treatment is appropriate for many, if not most, postmenopausal women, and it is comforting to know that progesterone-like hormones are effective in reducing the risk of womb cancer in these women. Many more studies on this combination therapy must be carried out with monitoring of side effects before it is completely accepted for most women. It also appears likely that simultaneous administration of extra calcium can complement the benefits of the hormonal therapy. For those women who are unsuitable for hormone treatment calcium supplementation may represent the only approach to prevention and treatment of osteoporosis (Drs R M Francis and P L Selby, 1987, *British Medical Journal*).

——Reversible bone loss in anorexia nervosa——

This is the title of a paper in the *British Medical Journal* (22 August 1987) by doctors from the Institute of Psychiatry, De Crespaigny Park, London and Guys Hospital, London. It reports complete recovery after osteoporosis in adults where the condition was induced by anorexia nervosa. Such individuals are well known to be susceptible to bone fractures including collapse of the spinal vertebrae (Dr N A Rigotti et al, *New England Journal of Medicine*, 1984, and Dr G I Szmukler et al, *British Medical Journal*, 1985).

Those with anorexia nervosa are likely to lose bone on two counts; the fact that they are oestrogen deficient, and their low intakes of calories and calcium because of their state of malnutrition. In order to determine the dominant factor in anorexia nervosa, four groups of subjects were studied.

The first group consisted of 45 patients with anorexia nervosa, aged 14 to 54 years, who had fallen to a minimum of 75 per cent of their weight before they developed the condition. All had had amenorrhoea for over one year. The second group were 31 normal volunteers in a similar age range, 19 to 46 years. Twenty-five patients who had recovered from anorexia nervosa (age range 25 to 52 years) comprised the third group. The final group consisted of matched individuals (age range 24–48 years) who were healthy. In every case, bone mineral density of the spine, the right hip, and the forearm was measured.

The results indicated that bone density was reduced in the anorectic patients in proportion to the duration of the illness. Twenty of these patients, who had been amenorrhoeic for six years or more, had the lowest bone density in the hip. What was significant, however, was that bone mineral density improved as the body weight of these patients increased. In addition, those in the third group (the patients who had recovered from anorexia nervosa) demonstrated normal bone mineral densities. Despite their low intakes of food the anorectic patients had normal blood values of calcium indicating that the bones were acting virtually as the sole supplier of the mineral. Female sex hormone levels were low in the blood of all those patients. Nine of the forty-five anorectics had suffered bone fractures when the study was undertaken indicating that bone density had fallen below the critical level.

The study thus confirmed that in anorexia nervosa there is significant loss of bone from the three sites studied, namely hip, spine, and forearm. What was not possible was to distinguish with certainty the osteoporosis due to malnutrition from that due to secondary oestrogen deficiency. Indirect evidence that body weight has an independent effect on inducing osteoporosis comes from the finding that there was no correlation between bone density and blood oestrogen levels. Also there was improvement in bone density as body weight increased, even before the restoration of menstruation. The authors recommend that oestrogen treatment should not be given to young anorectic women who will restore their bone density anyway as they return to normal eating habits and gain weight.

Anorexia nervosa can thus be regarded as yet another condition that can induce osteoporosis, alongside old age, the menopause, corticosteroid therapy, amenorrhoea from any cause, and excessive physical activity. In all of them increased calcium intakes will help improve the condition with or without any other appropriate therapy.

————The symptoms of osteoporosis————

In most cases an individual is between 50 and 70 years of age when osteoporosis is diagnosed. We have seen, too, that young women can also be afflicted under certain circumstances. The problem is that at whatever time it strikes, osteoporosis is unlikely to cause effects that are immediately apparent to the sufferer. Usually nothing happens until the bones become so weak that a sudden strain, bump, or fall causes a fracture. Then of course pain becomes severe and sufficient to contain physical activity.

Crush fractures of the spinal vertebrae can occur with or without causing pain. If pain is present, further clues to the presence of osteoporosis may be loss of height or curvature of the upper back. The type of pain is usually a chronic aching along the spine, or more often, pain from spasms in the muscles of the back may occur. With a partially collapsed spine, the muscles of the back must take a greater share of supporting the upper half of the body so that, periodically, they react accordingly.

X-ray examination, usually for some other purpose, is often the means of discovering osteoporosis. The technique can often reveal vertebrae that are crushed or broken, unbeknown to the sufferer. Usually the person has to lose about a quarter of his or her bone mass before the loss can be detected on a regular X-ray. Moreover, by the time that this extent of bone loss has occurred, the bones may be already susceptible to breaking. Conventional X-ray examination therefore has its limitations.

Fortunately, new techniques for measuring bone density, particularly when this is being reduced in the early stages, are now being used and developed. Often they are research tools only and so denied to many practising doctors. One technique is called 'photon absorptiometry' or 'photon densitometry' where rays, like X-rays, are passed through the skeleton and a machine measures how much of the ray is absorbed by the bone. The figures then give an indication of how dense the bones are since the denser the bones, the less rays will be detected beyond the skeleton.

There are also more sophisticated X-ray techniques to measure bone density. One is called computerized topography (CT) scanning. It depends on producing an image of a plane section across a solid object (in this case the bone) by superimposition of exposures at multiple angles. In this way a three-dimensional picture of bone density can be built up. This method has been suggested as suitable for mass bone mineral screening for osteoporosis (eg 'Sounding Board' *New England Journal of Medicine* 22 January 1987) but the economics of the techniques have given it low priority in most countries because of restricted health resources.

At present there are no blood or urine tests or assays that assist in the diagnosis of osteoporosis. Blood calcium levels are meaningless because the tiny amount of the mineral in the blood is always maintained by using the skeleton as a reservoir. Similarly phosphorus and vitamin D levels are invariably normal in osteoporosis. The various methods of measuring bone density therefore remain the only non-invasive diagnostic techniques available at present.

CHAPTER 2

Fulfilling your requirements

Calcium is an essential mineral for man and all animal and fish life, in fact anything that possesses a skeleton. It is also an essential component of shells. Calcium has the chemical symbol Ca with an atomic weight of 40.08. It is classed as an alkaline earth metal.

Calcium is the fifth most abundant mineral in the earth's crust (to the extent of 3.64 per cent) so it is widely distributed in soils. Its concentration amongst soils does however vary widely and this is reflected in the calcium content of the plants growing in these soils. Calcium Carbonate (or limestone) is the most common form of calcium found naturally, mainly as Portland Stone, dolomite, marble, and chalk. The sea contains substantial amounts of soluble calcium at levels up to 400g per 1,000kg seawater.

Water from rivers, springs, and wells contains calcium in small amounts depending on how acid the water is since it is this that dissolves the mineral from the rocks and soil through which the water flows. Hence hard water supplies not insignificant amounts of calcium and can be a useful dietary source of the mineral. Daily intakes of 250mg calcium from drinking water alone are quite possible.

In the western world it is now accepted that areas served by hard water usually have lower mortality rates from cardiovascular disease than areas where the water is soft. Hard water is less corrosive and is therefore believed to cause less leaching of potentially harmful metals such as lead, copper, and cadmium from the water-carrying pipes. Calcium can therefore be beneficial in this respect and its protective action is further evidence of its role in helping to rid the body of undesirable elements because it prevents the absorption and transfer of toxic materials from the intestine to the blood.

Hence, although the daily requirement of calcium is derived almost wholly from food, certain populations with calcium-poor diets may obtain a substantial amount of their needs from the water they drink. This is perhaps the case today in parts of India where the diet is lower in calcium than it is in western countries and where as much as one-fifth of calcium requirements can come from drinking water.

One must also remember that some bottled mineral waters can be useful sources of calcium. The contents of the mineral in some of the more popular brands are shown in Table 1. A water described as rich in mineral salts must contain at least 150mg calcium per litre.

Table 1 Calcium content of some popular mineral waters

Contrexeville	451.0
Ferrarelle	448.2
San Pellegrino	207.0
Badoit	157.0
Perrier	140.2
Bally Gower	121.2
Vichy Celestins	109.0
Ashbourne	102.0
Evian	78.0
Sainsbury Scottish Spring	50.0
Apollinaris	47.5
Highland Spring	44.8
Malvern	37.0
Vichy St Yorre	35.0
Brecon	25.0
Bru	23.3
Cwm Dale	13.1
Volvic	10.4
Spa Barisart	5.5
Spa Reine	3.5
Ramlosa	1.8

Soil is the ultimate source of all calcium and all growing plants will absorb it, albeit in varying amounts. All foods of vegetable origin thus contain small but useful quantities of the mineral unless they are highly refined and processed. Then, calcium is one of the first casualties and a moderate supplier of it becomes a poor one. Only in one case is the lost mineral added back to the basic food from which it was removed and that is all wheat flour, except wholemeal.

Animals concentrate the calcium from their vegetable diets in their milk which is why this food and products derived from it remain the richest source of the mineral. In communities who keep good dairy herds at least half their daily intake comes from milk and milk products. Those who drink little or no milk have relatively low calcium intake. As we shall see later, in most cases this is in no way detrimental.

In the average diet of the western world, milk, cheese, and yoghurt provide 56 per cent of the calcium intake. Liquid milk alone accounts for 36 per cent of our daily intake. A further 25 per cent comes from cereal foods, including bread, and another 7 per cent from vegetables. The calcium in the UK household food supply in 1984 has been calculated by the Ministry of Agriculture, Fisheries and Food to be supplied in the following way:

Table 2 Calcium in the food supply (1984)

	mg	% of total
Milk	309	35.8
Cheese	115	13.3
Other dairy foods	63	7.3
Total dairy foods	487	56.4
White bread	82	9.5
Other bread	41	4.7
Other cereal foods	95	11.1
	218	25.3
Total vegetables	58	6.7
Total meat and fish	42	4.9

These figures suggest an average calcium intake of 850mg which is only 93.3 per cent of the total. The shortfall is made up with sources that include nuts, fruits, and water.

Whilst this data applies to 1984, if we look at the contribution by individual foods to calcium intake from the UK household food supply in years previous to this, it is possible to detect a trend away from liquid milk as the main supplier of the mineral (see top of next page).

Wholemeal wheat flour contains an average of 35mg calcium per 100g (3.5 ounces) and unfortified white and brown flours contain only 15mg and 20mg respectively because during the refining process,

Table 3 Changes in calcium intakes over the years

Year	Milk	Cheese	All dairy	All cereal
1978	444mg	108mg	587mg	131mg
1981	400mg	117mg	561mg	118mg
1984	309mg	115mg	487mg	123mg
1978	45.0%	11.0%	59.5%	13.3%
1981	42.1%	12.4%	59.1%	12.4%
1984	35.8%	13.3%	56.4%	14.2%

calcium is lost. However, statutory fortification of white and brown flours with calcium increases their content of the mineral to 140 and 150mg respectively. Hence products made from these flours are better sources of calcium than those from wholemeal flour. This is reflected in the calcium contents of wholemeal, brown, and white bread which are 23,100, and 100mg per 100g. Flour-based foods made with the self-raising type of flour are rich in calcium because the mineral is a component of the commonly-used raising agents.

Of all the cereals, soya flour provides most calcium, with even more in the low fat variety than the full fat one. Some nuts too are rich in the mineral with almonds, brazils, barcelona, pistachio, and cola nuts holding pride of place. Vegetable foods vary tremendously in their calcium content but even those with lower values are useful suppliers of the mineral by virtue of the sheer bulk of their intake in the mixed diet. Seeds and pulses are excellent sources with sesame seed providing a massive 783mg calcium in every 100g, probably the richest natural calcium source available. Whilst high intakes of sesame seed are unlikely to be a feature of the average diet, other calcium-rich vegetables are. These include spinach, winged beans, parsley, watercress, kale, spring greens, spring onions, broccoli tops, and chick peas.

Most fish are useful providers of calcium but eating the flesh alone will give less of the mineral than if the whole fish is eaten as in the canned varieties because here the bones are softened and made edible by the canning process. Some primitive communities are known to grind both fish and animal bones into a form that can be eaten so providing themselves with calcium denied to them by their lack of intake of dairy produce. Shellfish are very good providers of calcium. On the whole, fish are richer sources of calcium than meat and poultry. This is because in animals and birds, most of the calcium is confined to the skeleton and the flesh has a very low concentration. Some offal supplies useful amounts as do certain meat products like sausages and pies but here much of the calcium is supplied by the cereal content.

A comprehensive list of foods and food products with their calcium levels is provided in Table 4. Kilocalorie and kilojoule levels are also given for the benefit of the calorie conscious reader. Most slimming regimes cut down on milk, dairy products, and cereals and so deny the slimmer his or her needs for calcium. Knowing both the mineral and energy contents of foods will enable the slimmer to work out a suitable diet that is not deficient in calcium. This particularly applies to the premenopausal female who tends to put on weight because of her age and attempts to correct this tendency with low-calcium slimming regimes.

Table 4 Calcium and energy contents of food items, products and prepared dishes as eaten (mg per 100g food)

Type of cheese	mg Calcium	Kcal	Kjoule
Parmesan	1290	379	1584
Emmental	1020	383	1604
Gruyere	1000	411	1720
Provolone	881	366	1532
Tilsit (45% fat)	858	356	1488
Tilsit (30% fat)	830	274	1145
Butter cheese	694	343	1436
Gouda	820	365	1529
Cheddar	810	391	1637
Cheshire	800	406	1682
Edam (30% fat)	800	255	1067
Edam (40% fat)	793	317	1326
Edam (45% fat)	678	355	1483
Roquefort	662	361	1509
Gorgonzola	612	357	1494
Bel Paese	604	374	1564
Camembert (30% fat)	600	219	916
Camembert (40% fat)	570	276	1156
Camembert (45% fat)	570	286	1195
Camembert (50% fat)	510	313	1311
Camembert (60% fat)	400	375	1570
Danish Blue	580	355	1471
Cheese spread	510	283	1173

Type of cheese	mg Calcium	Kcal	Kjoule
Limburger (20% fat)	510	188	787
Limburger (40% fat)	534	269	1124
Feta	429	248	1037
Mozzarella	403	226	947
Brie	400	342	1429
Stilton	360	462	1915
Processed (45% fat)	547	269	1125
Ricotta	274	172	721
Muenster (50% fat)	230	320	1340
Muenster (45% fat)	310	291	1217
Quark (skimmed milk)	92	76	317
Quark (20% fat)	85	112	468
Quark (40% fat)	95	161	673
Romadur (20% fat)	448	182	764
Romadur (30% fat)	374	222	927
Romadur (49% fat)	403	275	1152
Romadur (45% fat)	273	293	1228
Romadur (50% fat)	264	311	1302
Acid Curd (10% fat)	125	134	561
Blue (50% fat)	526	352	1472
Layered (10% fat)	77	75	316
Layered (20% fat)	79	94	395
Layered (40% fat)	82	145	608
Processed (60% fat)	355	323	1353
Cottage	95	105	438
Fresh (60–85% fat)	79	325	1360
Fresh (50% fat)	98	266	1114

Type of Milk	mg Calcium	Kcal	Kjoule
Dried skimmed	1290	366	1529
Dried whole	920	490	2051
Buttermilk powder	894	381	1596
Dried whey	890	351	1468
Condensed skimmed	340	271	1135
Evaporated whole	280	158	660
Condensed whole	242	132	554
Buffalo	195	107	449
Sheep	183	96	402
Camel	132	77	322

Type of Milk	mg Calcium	Kcal	Kjoule
Goat	130	71	296
Cow, skimmed	123	35	145
Cow, whole, raw	120	66	278
Cow, whole, pasteurized	120	65	271
Cow, whole, sterilized	120	47	198
Cow, whole, UHT	120	66	276
Cow, whole reduced fat	118	48	200
Mare	110	47	198
Ass	110	42	175
Buttermilk	109	35	148
Sour cream	100	184	769
Sterilized, canned	80	230	950
Whey	68	24	102
Human, transitional 6–10 days	40	64	269
Human	31	69	287
Human, transitional 2–3rd day	29	56	235
Single cream (21% fat)	79	212	876
Double cream (48% fat)	50	447	1841
Whipping cream (35% fat)	63	332	1367
Yogurt (fruit flavour)	160	95	405
Yogurt (nut flavour)	160	106	449
Yogurt (low fat)	143	39	163
Yogurt (3.5% fat)	120	69	289
Yogurt (1.5% fat)	114	50	209
Butter	15	740	3041
Margarine	4	730	3000
Soya milks S. Formula	56	67	280
Prosobee	60	67	280
Wysoy	64	67	280

Eggs	mg Calcium	Kcal	Kjoule
Chicken (dried yolk)	282	682	2853
Chicken (dried whole)	190	588	2459
Chicken (yolk)	140	359	1504

Eggs	mg Calcium	Kcal	Kjoule
Chicken (dried white)	84	371	1552
Duck (whole)	63	190	793
Chicken (whole)	56	160	670
Chicken (white)	11	53	222
Fried chicken egg	64	232	961
Scrambled chicken egg	60	246	1018
Boiled chicken egg	52	147	612
Poached chicken egg	52	155	644
Omelette	47	190	787

Egg and cheese dishes	mg Calcium	Kcal	Kjoule
Welsh rarebit	420	365	1523
Quiche lorraine	260	391	1627
Pizza	240	234	982
Cheese pudding	230	170	707
Cheese souffle	230	252	1049
Macaroni cheese	180	174	726
Cauliflower cheese	160	113	471
Scotch egg	56	279	1159

Fish	mg calcium	Kcal	Kjoule
Whitebait, fried	860	525	2174
Sprats, fried	710	441	1826
Sardines (canned in oil, fish only)	550	217	906
Sardines (canned, fish plus oil)	460	334	1382
Sardines (canned, fish plus tomato sauce)	460	177	740
Shrimps (boiled)	320	117	493
Pilchards (canned)	300	126	531
Mussels (boiled)	200	87	366
Oysters (raw)	190	51	217
Prawns (boiled)	150	107	451
Winkles (boiled)	140	74	312
Cockles (boiled)	130	48	203
Bloaters (grilled)	120	251	1043
Crab (canned)	120	81	341

Sardines & much better the salmon!

Calcium

Fish	mg calcium	Kcal	Kjoule
Scallops (steamed)	120	105	446
Shrimps (canned)	110	94	398
Haddock (fried)	110	174	729
Scampi (fried) ·	99	316	1321
Lemon sole (fried)	95	216	904
Plaice (fried)	93	279	1165
Salmon (canned)	93	155	649
Sardines (freshly cooked)	85	130	542
Cod (fresh battered)	80	199	834
Haddock (steamed, smoked)	58	101	429
Haddock (steamed)	55	98	417
Kipper (baked)	65	205	855
Skate (fried)	50	199	830
Lobster (boiled)	62	119	502
Whelks (boiled)	54	91	385

All other fish provide less than 50mg calcium

Cod (poached)	29	94	396
Cod (baked)	22	96	408
Cod (steamed)	15	83	350
Cod (grilled)	10	95	402
Halibut (steamed)	13	131	553
Lemon sole (steamed)	21	91	384
Plaice (steamed)	38	93	392
Saithe (steamed)	19	99	418
Whiting (fried)	48	191	801
Whiting (steamed)	42	92	389
Eel (stewed)	21	201	839
Herring (fried)	39	234	975
Herring (grilled)	33	199	828
Mackerel (fried)	28	188	784
Salmon (steamed)	29	197	823
Salmon (smoked)	19	142	598
Trout (steamed)	36	135	566
Dogfish (fried)	42	265	1103
Crab (boiled)	29	127	534
Tuna	40	232	969

Fish products	mg calcium	Kcal	Kjoule
Fish cakes (fried)	70	188	785
Fish fingers (fried)	45	233	975
Fish paste	280	169	704
Fish pie	40	128	540
Kedgeree	36	151	633
Roe (fried, hard)	17	202	844
Roe (fried, soft)	16	244	1019

Vegetables	mg calcium	Kcal	Kjoule
Sesame seed	783	–	–
Spinach (boiled)	600	30	128
Winged beans	530	–	–
Parsley (raw)	330	21	88
Soya beans	257	304	1272
Watercress (raw)	220	14	61
Kale	212	23	95
Dandelion leaves	158	–	–
Spring onions (raw)	140	35	151
Chives	129	–	–
Pigeon peas	121	270	1129
Horseradish (raw)	120	59	253
Chick peas	110	292	1220
Fennel leaves	109	–	–
Marigold	103	–	–
Cowpeas	101	255	1068
Broccoli tops (raw)	100	23	96
Sunflower seeds	98	535	2240
Turnip tops	98	11	48
Purslane	95	–	–
Spring greens (boiled)	86	10	43
Okra	70	17	71
Kohlrabi	68	22	92
Chick peas (cooked)	67	144	610
Mustard & Cress	66	10	47
Haricot beans	65	93	396
Onions (fried)	61	345	1424
Leeks (boiled)	61	24	104
Turnips (boiled)	55	14	60
Savoy cabbage (boiled)	53	9	40

Calcium

Vegetables	mg calcium	Kcal	Kjoule
Celery (raw)	52	8	36
Celery (boiled)	52	5	21

All other vegetables contain less than 50mg calcium

	mg calcium	Kcal	Kjoule
Ackee (canned)	35	151	625
Artichokes (boiled)	30	18	78
Asparagus (boiled)	26	18	75
Aubergine (raw)	10	14	62
Beans (French, boiled)	39	7	31
Beans (runner, boiled)	22	19	83
Beans (broad, boiled)	21	48	206
Beans (butter, boiled)	19	95	405
Beans (baked in tomato sauce)	45	64	270
Beansprouts (canned)	13	9	40
Beetroot (boiled)	30	44	189
Brussel sprouts (boiled)	25	18	75
Cabbage (spring, boiled)	30	7	32
Cabbage (winter, boiled)	38	15	66
Carrots (old, boiled)	37	23	98
Carrots (young, boiled)	29	20	87
Carrots (canned)	27	19	82
Carrot (juice)	27	21	88
Cassava	37	132	554
Cauliflower (raw)	21	13	56
Cauliflower (boiled)	18	9	40
Celeriac (boiled)	47	14	59
Chicory (raw)	18	9	38
Cucumber (raw)	23	10	43
Endive (raw)	44	11	47
Garlic	38	–	–
Laverbread	20	52	217
Lentils (cooked)	13	99	420
Lettuce	23	12	51
Marrow (boiled)	14	7	29
Mushrooms (raw)	3	13	53
Mushrooms (fried)	4	210	863
Onions (raw)	31	23	99
Onions (boiled)	24	13	53

Vegetables	mg calcium	Kcal	Kjoule
Parsnips (boiled)	36	56	238
Peas (fresh, boiled)	13	52	223
Peas (frozen, boiled)	31	41	175
Peas (canned)	24	47	201
Peas (processed)	27	80	339
Peas (dry boiled)	24	103	438
Peas (split, boiled)	11	118	503
Peppers (green, raw)	9	15	65
Peppers (green, boiled)	9	14	59
Plantain (green, raw)	7	112	477
Plantain (green, boiled)	9	122	518
Plantain (fried)	6	267	1126
Potatoes (boiled)	4	80	343
Potatoes (mashed)	12	119	499
Potatoes (baked with skins)	9	85	364
Potatoes (roast)	10	157	662
Potato (chips)	14	253	1065
Potatoes (canned)	11	53	226
Potato (reconstituted powder)	20	70	299
Potato (crisps)	37	533	2224
Pumpkin (raw)	39	15	65
Radishes (raw)	44	15	62
Seakale (boiled)	48	8	33
Swedes (boiled)	42	18	76
Sweetcorn (boiled)	4	123	520
Sweetcorn (canned)	3	76	325
Sweet potatoes (boiled)	21	85	363
Taro	31	104	434
Tomatoes (raw)	11	14	60
Tomatoes (fried)	13	69	288
Tomatoes (canned)	11	12	51
Yam (boiled)	9	119	508

Nuts	mg calcium	Kcal	Kjoule
Almonds	250	565	2336
Barcelona	170	639	2637
Brazil	180	619	2545

Nuts	mg calcium	Kcal	Kjoule
Pistachio	136	554	2318
Cola nut	108	227	948
Pecan	73	–	–
Peanuts (fresh)	61	570	2364
Peanuts (roasted)	61	570	2364
Walnuts	61	525	2166
Macadamia	51	–	–
Chestnuts	46	170	720
Hazel (cob) nuts	44	380	1570
Peanut butter	37	623	2581
Cashews	31	536	2242

Cereals	mg calcium	Kcal	Kjoule
Rich sources: more than 100mg per 100g edible cereal			
Wheat bran	110	206	872
Wheat flour, brown (85%)	150	327	1392
Wheat flour, white breadmaking	140	337	1433
Wheat flour, white household	150	350	1493
Wheat flour, self-raising	350	339	1443
Wheat flour, patent (40%)	110	347	1480
Soya flour, full fat	210	447	1871
Soya flour, low fat	240	352	1488
Bread, brown	100	223	948
Bread, white	100	233	991
Bread, soda	150	264	1122
Bread rolls, all types	120	300	1290
Muesli	200	368	1556
Biscuits, chocolate	110	524	2197
Biscuits, cream crackers	110	440	1857
Biscuits, digestive, plain	110	471	1981
Biscuits, ginger nuts	130	456	1923

Cereals	mg calcium	Kcal	Kjoule
Biscuits, sandwich	100	513	2151
Biscuits, semi-sweet	120	457	1925
Biscuits, water	120	440	1859
Gingerbread	210	373	1573
Rock cakes	390	394	1658
Sponge cake	140	464	1941
Shortcrust pastry	110	527	2202
Scones	620	371	1562
Pretzels	147	283	1193
Scotch pancakes	120	283	1193
Bread & butter pudding	130	159	668
Egg custard	130	118	497
Custard tart	110	287	1199
Dumpling	160	211	885
Dairy ice-cream	140	167	704
Non-dairy ice-cream	120	165	691
Milk pudding	130	131	552
Yorkshire pudding	130	215	902
Pancakes	120	307	1286

Moderate sources	mg calcium	Kcal	Kjoule
Between 50 and 100mg calcium per 100g edible food			
Wheatgerm	69	285	1191
Oatmeal	55	401	1698
Fried bread	90	558	2326
Currant bread	90	250	1063
Malt bread	94	248	1054
Chapatis	66	336	1415
All-bran	74	273	1156
Ready Brek	64	390	1651
Crispbread, rye	50	321	1367
Crispbread, wheat	60	388	1642
Chocolate digestive biscuits	84	493	2071
Oatcakes	54	441	1855
Short-sweet biscuits	87	469	1966
Shortbread	97	504	2115
Wafers, filled	73	535	2242

Moderate sources	mg calcium	Kcal	Kjoule
Fruit cake, all types	75	332	1403
Currant buns	90	302	1279
Doughnuts	70	349	1467
Jam tarts	62	384	1616
Mince pies	76	435	1826
Flaky pastry	90	565	2356
Cheesecake	67	421	1747
Christmas pudding	87	304	1279
Fruit pies	51	369	1554
Canned rice pudding	93	91	386
Treacle tart	65	371	1563
Trifle	82	160	674
Oats	80	348	1455

Other sources	mg calcium	Kcal	Kjoule
Less than 50mg calcium per 100g edible food			
Pearl barley	14	297	1243
Barley	38	286	1197
Cornflour	15	354	1508
Macaroni	8	117	499
Porridge	6	44	188
Maize	15	318	1329
Rice, unpolished	23	342	1431
Sorghum	26	303	1268
Rye bread	23	195	818
Pasta	27	343	1433
Rusks	42	378	1584
Rice, polished	6	89	371
Rice flour	7	349	1462
Rye flour	32	335	1428
Sago	10	355	1515
Semolina	18	350	1489
Spaghetti, boiled	7	117	499
Spaghetti, canned	21	59	250
Tapioca	8	359	1531
Wholemeal bread	23	216	918
Cornflakes	3	368	1567

Other sources	mg calcium	Kcal	Kjoule
Grapenuts	37	355	1510
Puffed wheat	26	325	1386
Rice krispies	7	372	1584
Shredded wheat	38	324	1378
Special K	42	388	1650
Sugar puffs	14	348	1482
Weetabix	33	340	1444
Madeira cake	42	393	1652
Eclairs	48	376	1569
Apple crumble	28	208	878
Lemon meringue pie	46	323	1359
Meringues	4	380	1620
Buckwheat	21	334	1399

Meats	mg calcium
Bacon	7–16
Beef	7–18
Lamb	7–11
Pork	7–11
Veal	8–14
Chicken	9–12
Duck	14–20
Goose	10
Grouse	30
Partridge	46
Pheasant	49
Pigeon	16
Turkey	8–12
Hare	21
Rabbit	11
Venison	29
Brain	11–16
Heart	5–10
Kidney	8–16
Liver	6–15
Oxtail	14
Sweetbread	34
Tongue	11–31

Meats	mg calcium
Tripe, stewed	150
Canned meats	9–32
Offal products	26–55
Sausages	23–73
Pies & pastries	47–82
Cooked meat dishes	12–110

Fruits	mg calcium
Apples	3–4
Apricots, raw and cooked	13–17
Apricots, dried	92
Apricots, dried and stewed	34
Apricots canned	12
Avocado pears	15
Banana	7
Bilberries	10
Blackberries, raw	63
Blackberries, stewed	54
Cherries, raw	16
Cherries, stewed	18
Cranberries	15
Currants, black, raw	60
Currants, black, stewed	51
Currants, red, raw	36
Currants, red, stewed	31
Currants, white, raw	22
Currants, white, stewed	19
Currants, dried	95
Damsons, raw	24
Damsons, stewed	20
Dates, dried	68
Figs, green, raw	34
Figs, dried, raw	280
Figs, dried, stewed	160
Fruit pie filling, canned	18
Fruit salad, canned	8
Gooseberries, green, raw	28

Fruits	*mg* *calcium*
Gooseberries, green, stewed	24
Gooseberries, ripe, raw	19
Grapes, black	4
Grapes, white	19
Grapefruit, fresh	17
Grapefruit, canned	17
Greengages, raw	17
Greengages, stewed	15
Guavas, canned	8
Lemons	110
Lemon juice	8
Loganberries, raw	35
Loganberries, stewed	32
Loganberries, canned	18
Lychees, raw	8
Lychees, canned	4
Mandarins, canned	18
Mangoes, canned	10
Medlars, raw	30
Melons, cantaloupe	19
Melons, honeydew	14
Melons, water	5
Mulberries	36
Nectarines	4
Chives	61
Oranges	41
Oranges juice	12
Passion fruit	16
Paw Paw, canned	23
Peaches, raw	5
Peaches dried, raw	36
Peaches dried, stewed	13
Peaches, canned	4
Pears, raw	6
Pears, stewed	6
Pears, canned	5
Pineapple, fresh	12

Fruits	*mg calcium*
Pineapple, canned	13
Plums	11–14
Prunes, raw	38
Prunes, stewed	18
Quinces	14
Raisins	61
Raspberries, raw	41
Raspberries, stewed	43
Raspberries, canned	14
Rhubarb, raw	100
Rhubarb, stewed	93
Strawberries, raw	22
Strawberries, canned	14
Sultanas	52
Tangerines	42

Miscellaneous foods	*mg calcium*
Demerara sugar	53
Golden syrup	26
Black treacle	500
Marmalade	120
Mincemeat	30
Chocolate, milk	220
Chocolate, plain	38
Chocolate, filled	92
Fruit gum	360
Toffees	95
Cocoa powder	130
Coffee, ground	130
Coffee, instant	160
Coffee, drink	2
Drinking chocolate	33
Horlicks	230
Ovaltine	36
Baking powder	11,300
Bovril	40
Curry powder	640

Miscellaneous foods	*mg calcium*
Ginger	97
Marmite	95
Oxo cubes	180
Mustard powder	330
Pepper	130
Block salt	230
Table salt	29
Dried yeast	80
Compressed baker's yeast	25

Vegetarians who eat dairy products need have no fear about their dietary calcium intake since denying themselves meat, fish, and poultry will have a minimum effect on calcium because of its relatively low content in these foods. Vegans, who partake of no food of animal, fish or fowl origin, including milk and milk products, can easily obtain their calcium needs from cereals, nuts, fruit, and vegetables as long as they are aware of the richer sources of the mineral amongst the foods they do eat.

Recommended dietary allowances ——————(RDA) of calcium——————

The recommended dietary allowances of nutrients are defined in the latest US RDA report (Committee on Dietary Allowances, Food and Nutrition Board, 1980, *Recommended Dietary Allowances,* 9th revised edition, National Academy of Sciences, Washington, DC). The report states:

> RDA are recommendations for the average daily amounts of nutrients that population groups should consume over a period of time. RDA should not be confused with requirements for a specific individual. Differences in the nutrient requirements of individuals are ordinarily unknown.

The current British recommendations report (DHSS, 1979, *Recommended Daily Amounts of Food Energy and Nutrients for Groups of People in the United Kingdom;* Report by the Committee on Medical Aspects of Food Policy; Report on Health and Social Subjects 15. HM Stationery Office, London) states that since publication of the 1969 report 'more difficulties have been encountered about the use of figures than about their validity. Although these figures were intended to apply to groups of people, they have been used mistakenly as recommendations for individuals'. It goes on to say that:

Experience has shown that the distribution of nutrient intakes in a group of healthy people is such that many individuals eat less than the amounts put forward in the 1969 recommendations (DHSS, Reports on Public Health and Medical Subjects, No 120, 1969) without any recognizable signs of deficiency.

A more practical definition of the recommended amount of a nutrient is as follows:

The average amount of the nutrient which should be provided per head in a group of people if the needs of practically all members of the group are to be met.

A third definition is given by the *FAO/WHO Handbook on Human Nutritional Requirements,* WHO Monograph series No 61, WHO, Geneva. This says that:

The figures for recommended intakes may be compared with actual consumption figures determined by food consumption surveys. Such comparisons though always useful, cannot in themselves justify statements that undernutrition, malnutrition or overweight is present in a community or group, as such conclusions must always be supported by clinical or biochemical evidence. The recommended intakes are not an adequate yardstick for assessing health because each figure represents an average requirement augmented by a factor that takes into account inter-individual variability.

What do these definitions mean to the individual who wishes to assess the adequacy of his or her calcium intake in the diet? The UK report (1979) is quite candid. It admits that:

Recommended amounts have a limited use in evaluation of the results of surveys of the amounts of food eaten by individuals. Since the distribution of requirements for nutrients is not known, it is not possible to estimate the probability that an individual is undernourished by comparing his or her intake with the recommended amount. Nevertheless, it would still be true to say that, on present knowledge, the greater the possibility that some individuals may be undernourished with respect to the nutrient in question!

The US Report (1980) also has something to say about the individual intakes. It states that:

> Even if a specific individual habitually consumes less than the recommended amounts of some nutrients, his diet is not necessarily inadequate for those nutrients. However, since the requirements of each individual are not known, it is clear that the more habitual intake falls below the RDA and the longer the low intake continues, the greater is the risk of deficiency.

A clue to why these official bodies differ in their definition of RDAs is to be found in such figures as estimated. There are several approaches to the problem of estimating requirements. Preliminary information comes from observations on the range of intakes by apparently healthy people; in some instances it is supplied by results from experimental deficiency in animals. The four principal direct types of information are:

(i) On how long an intake of a nutrient (eg calcium) does deficiency disease start to occur?

(ii) How little of the nutrient will cure clinical signs of deficiency observed in spontaneous disease or induced experimentally in volunteers?

(iii) What is the lowest intake of the nutrient that maintains metabolic balance over a long period? This approach is the most used in assessing calcium needs.

(iv) What is the minimum intake needed for tissue saturation or to give normal function tests for the nutrient? This approach is more use in determining vitamin requirements but may be applicable to calcium also.

We should expect that if all countries used these criteria, and presumably all have access to literature information to back up their own data, the calculated daily allowances of nutrients such as calcium would be similar. They are not, as Table 5 shows. For calcium, one important reason is the possibility of adaptation to low intakes by generations of the population who eat them. For example, European adults are usually in negative balance (ie less goes in than comes out) at intakes below 800mg calcium per day while many Africans and Asians maintain healthy bones on only 500mg per day. As we shall see later, there are many studies indicating that some communities maintain a positive calcium balance (ie more goes in than comes out) on intakes far below 500mg daily. The ideal situation is where calcium intake equals calcium output.

According to Dr R M Leverton ('The RDAs are for amateurs.' *Journal of the American Dietetic Association,* 1975), RDAs can be minimized and often they are expected to provide more or different information than they were ever designed to give. Most recommendations by the different authorities:

(i)   Are for nutrients as eaten after food processing and cooking.

(ii)   Are for healthy people. They do not allow for illness or major stresses in life.

(iii) May be affected by a variety of drugs.

(iv)   Are more than enough for most people. They are therefore too high themselves as criteria for inadequate food intake.

(v)   Do not have to be eaten every single day; a low intake one day can be balanced the next day.

(vi)   Do not say that you can't eat more of the nutrient than the recommended amount but do not indicate at what higher level toxic effects might start.

(vii) Assume a certain nutritive quality, biological value of availability in the body. For example, the recommendation for calcium will assume only a certain proportion is absorbed.

(viii) Assume that enough of other major nutrients is absorbed. For example, if vitamin D is not present, calcium may not be absorbed.

(ix)   Are for standard body size (eg weight) and range of usual exercise, usually defined as sedentary, moderately active, or very active.

(x)   Are for oral intakes of the usual foods of the country.

(xi)   Are for major nutrients only. It is assumed that if the major nutrients are adequate then so are the minor ones that accompany them.

(xii) Cannot fully allow for a depletion that can occur to high or low intakes of, for example, calcium.

(xiii) Do not allow for interactions between nutrients eg calcium and phytic acid.

Let us now look at the reasoning behind the specific RDAs for calcium suggested by the UK and USA authorities. As far as the UK is concerned, the recommendations in the 1979 report for adults are the same as those in the 1969 report. Hence it is possible to quote the 1969 report as the arguments in it are still considered valid. The only difference in the two reports is an increased recommendation of calcium for children between one and nine years to 600mg daily. This is the same as that for those aged between birth and one year.

Originally, the British Medical Association (1950) recommended 800mg calcium a day for adults after consideration of balance studies made chiefly in western countries where the intake of the mineral is high. However it was later realized that adaptation to a lower calcium intake occurs after a period of time and the negative calcium balance shown at first on reducing the calcium intake below 800mg does not necessarily mean that the body requires this amount of calcium. In 1962 the FAO/WHO experts suggested a 'practical allowance' for adults of 400 to 500mg calcium daily so the DHSS Panel on Recommended Allowances of Nutrients also suggested this amount as there appeared to be no evidence of calcium deficiency in countries where calcium intakes were of this order.

Other arguments for accepting this allowance came from a consideration of the evidence available up to 1969 on the relationship between calcium intake and osteoporosis. The same FAO/WHO report quoted above (*Calcium requirements; report of an FAO/WHO Expert Group,* FAO Nutr Mtg Rep Sep 30 and Tech Rep Ser World Health Organisation 230) considered that the evidence supporting the view that osteoporosis may be due to an inadequate dietary supply of calcium was not convincing. This evidence was still accepted ten years later in the latest RDA figures published by the DHSS so it is worth reviewing it here.

Factors other than diet

Dr B E C Nordin in the *Journal Clinical Orthopaedics* of 1966 surveyed the international prevalence of osteoporosis and concluded that it occurred in old people regardless of diet but it did not appear in young people whatever their calcium intake. There is no doubt that osteoporosis develops more quickly in women than in men, and the condition becomes increasingly common with age in both sexes. The American researchers S M Garn, G C Rothman and B Wagner in their 1967 study reported in *Federation Proceedings,* Federation of American Society of Experimental Biology, that there was no relationship between bone loss and calcium intake. They thus concluded that, beginning by the fifth decade in both sexes, bone loss is a general phenomenon, and progresses more than twice as fast in the female as in the male. They stated that 'intakes of calcium above 1,500mg a day do not seem to be protective, and levels of calcium intake even below

300mg are not demonstrably associated with bone loss'.

In the same issue of the journal, Dr R W Smith found that some women lost bone despite diets rich in calcium, whereas others of similar heights and weights apparently lost no more bone through years of very low consumption of calcium. He concluded that some factor or factors other than diet were the cause of this seemingly normal or inherent bone loss. As we have seen in Chapter 1, one of these factors appears to be the female sex hormones, and in the case of men, male sex hormones.

In the *Lancet* of 1968, Drs H F Newton-John and D B Morgan suggested that the amount of bone present in old age is related to the amount in early adult life and not to subsequent calcium intake. This idea does seem to have been confirmed in later studies (as reported in Chapter 1), but the age at which bones can be strengthened by extra calcium intake is now accepted as later than early adult life.

On the basis of this evidence, the 1979 recommendations by the Committee on Medical Aspects of Food Policy for calcium intakes are the same as those of the 1969 committee. In the words of this committee 'In view of the evidence that it is impossible to prevent osteoporosis with dietary calcium in adult life, there is no reason, if vitamin D is adequate, for departing from the FAO/WHO recommendations for calcium intake for adults.' In view of the present epidemic of osteoporosis in the UK, it will be interesting to see if the same conclusions are reached by the new committee when their recommendations eventually come out.

Let us now look at the reasoning behind the suggested intakes for the UK population apart from non-pregnant adults. Children have their own recommended intakes of calcium and the first ones were based on the fact that families with three or more children fail on average to reach the British Medical Association (1950) recommended level of the mineral intake. This has been the position for many years, although the average intake has not fallen to less than 80 per cent of the recommended level, the lower limit of the range has been taken as a signal for 'watchful concern' (MAFF, *Household Food Consumption and Expenditure,* 1965). Although National Food Survey findings have been interpreted by some people to mean that these families are calcium deficient, no clinical evidence of primary calcium deficiency has been reported in the UK or indeed anywhere in the world.

It is known that children in large families in the UK are smaller than their counterparts in small families but this might be due to a number of factors, not all of them nutritional. There is no evidence that children in countries where the calcium intake is low fail to reach full height because they are calcium deficient; addition of calcium to the diet has not resulted in improved growth.

The amount of bone possessed in early adult life is determined by genetic factors and physical activity and possibly also by calcium intake in childhood. The human adult body contains 1,200g of calcium so if this is to be provided in 20 years the average amount of calcium required to be retained during this time is 160mg per day although the amount varies with rate of growth and body size. It is impossible to infer the required dietary intake of calcium from the tissue demand, until the extent of the variability in the efficiency of alimentary absorption is established.

In 1950, the British Medical Association recommended 1g of calcium per day for children and 1.1 to 1.4g per day for different ages and sexes of adolescents but later it was felt that those intakes may be in excess of requirements. There is, however, justification if a high calcium intake during the growing period is needed for maximum calcification of the skeleton. Similarly if maximum calcification of the skeleton in early adult life leads to relatively large amounts of bone in old age then such high intakes are justified. Although the committee of the time could not produce or find concrete evidence for these possibilities, until such evidence does become available it would be a wise precaution to ensure your children do have more rather than less of the recommended intakes.

There is one concession in the 1979 figures when compared to those suggested by the 1969 committee. It is recognized that between one and nine years of age, the bones of children show a greater increase in the concentration of calcium than at any other time during the growth period. The suggested intake has therefore been increased from 500 to 600mg daily for children in this age range to make it the same as that recommended between birth and one year.

In women who are pregnant the additional calcium required for the foetus is about 30g; in those who are breast-feeding, between 150 and 300mg calcium is secreted daily in the milk during this period. Part of the extra requirement may be met by increased intestinal absorption and part from the skeleton. To help overcome this drain on the bones the recommended intake of calcium is increased to 1,000–1,200mg per day during the last third of pregnancy and throughout lactation. These figures, first suggested by FAO/WHO in 1962, are still accepted today by the UK committee on RDAs.

Let us now look at the US RDAs as decided by the Committee on

Dietary Allowances, Food and Nutrition Board 1980 *Recommended Dietary Allowances,* 9th revised edition, National Academy of Sciences, Washington DC. The RDAs for calcium have not changed between the 1974 and 1980 suggestions of the committees so conclusions reached in the 1974 report have been accepted in the later report. Hence it was recommended that 800mg calcium should be consumed daily by the adult, on the basis that 320mg are lost daily by various routes and that 40 per cent of dietary calcium is absorbed. Based on these figures, 800mg of calcium is required in the daily diet just to maintain equilibrium. These recommendations, based on the studies by Dr G A Goldsmith and reported in a symposium on *Food and Civilisation* (1966), published by C C Thomas, Springfield, Illinois, were a result of experiments conducted on groups of individuals accustomed to ample intakes of foods high in calcium as in western diets. Later reports suggested that adults remain in calcium balance despite lower calcium intakes — the same conclusion reached by the FAO/WHO report (1962) referred to above.

These studies have shown that men adapt, with time, to lower calcium intakes and maintain calcium balance on intakes as low as 200–400mg per day. Furthermore,* a higher proportion of calcium is utilized when intake is low than when it is liberal. However, most of the national groups cited as being in equilibrium on such low calcium intakes either live in tropical or semi-tropical areas with abundant sunlight or may well have unrecognized sources of calcium in the diet. For these reasons, the US authorities consider it unwise to recommend such low calcium intakes and prefer the higher and safer figure of 800mg.

Quoting the same evidence that the UK authorities did in the discussion above, the USA group reach a different conclusion. Whilst admitting that osteoporosis is not a disease that comes on suddenly in middle or old age and that it is not preventable by increasing calcium intake, the US experts do point out that work by Dr B E C Nordin in the *American Journal of Clinical Nutrition* (1962) showed that calcium supplements can induce calcium retention and relieve the symptoms of osteoporosis. This finding may reflect the fact that although the efficiency of absorption decreases with the amount of calcium in the diet, the total quantity of calcium retained actually increases. It is also important to remember that high protein intakes accompanying calcium will give rise to excessive calcium losses. Prolonged high protein–low calcium diets will therefore eventually lead to significant calcium, and hence bone, losses.

Taking all these points into consideration, the US recommend that for planning food supplies and as a guide for the interpretation of food consumption of groups of adults, 800mg per day of calcium is recommended. Those who eat less than the customary United States intake of protein will remain in calcium balance with intakes below the allowance recommended.

Like all authorities, those in the US recommend increased intakes of the mineral during pregnancy and whilst breast-feeding. These are required even though there is increased efficiency of intestinal absorption during these states. There is also evidence that calcium may be stored by the mother in excess of the needs of the foetus and this extra mineral will stand her in good stead during subsequent breast-feeding. Hence based on the fact that the full term infant contains about 25g of calcium and that during lactation a woman will lose between 250 and 300mg calcium per day in the milk, the recommended daily intake during both periods is set at 1,200mg. If a woman has a high production of milk she can lose almost 1,000mg calcium daily via the milk so calcium needs may be even higher than the average 1,200mg recommended and should be adjusted.

A breast-fed infant receives about 300mg calcium per litre of milk and retains about two-thirds of it. In contrast, an infant fed the usual cow's milk formula containing added carbohydrate will receive about 170mg of calcium per Kg body weight (compared to 60mg in the breast-fed infant) but will retain only between 25 and 30 per cent. Whilst the breast-fed infant has less calcium fed to it, its calcium needs are fully met by breast-feeding because of more efficient assimilation. The allowances for the first year of life apply only to the infants fed formula milk.

The US allowance of 360mg calcium daily in the first six months of life and 540mg during the next six are below the 600mg recommended by the UK authorities. However, after this age there is a wide divergence because the US suggest 800mg daily for all children. The reason that this is the same quantity as an adult is because in terms of intake per unit of weight (pounds or kilos), growing children need between two and four times as much calcium as does an adult. Children weigh less, therefore their calcium intake is proportionally higher. At the same time, the actual retention of calcium by healthy children is higher than that of an adult.

Again, in contrast to the UK recommendations, the US experts accept that the rapid growth rate in children that characterizes pre-adolescence and puberty (10 to 18 years) calls for higher intakes of

68 *Calcium*

calcium at these ages. Therefore the higher intake of 1,200mg daily is recommended. At this level in the diet, children have shown maximum calcium retention (Dr L G J Bogart and colleagues, 1966, in *Nutrition and Physical Fitness,* 8th edition, W B Saunders Co, Philadelphia, Pa, USA).

Look to the higher
————————intake recommendations————————

Both UK and US recommendations are under review at present and it is possible that both will come closer together as figures are revised. However, until this happens, the higher recommended allowances of the US authorities would be the more sensible to try and achieve. There is a growing belief amongst nutritionists of many nations that it is better to err on the side of higher rather than lower intakes of calcium.

The recommended dietary intakes around the world are shown in Table 5. Most of the figures are taken from 'Nutrition Abstracts and Review', reviews in *Clinical Nutrition,* Vol. 53, No. 11, November 1983, and form part of a report by Committee 1/5 of the International Union of Nutritional Sciences (1982).

The table shows the most recent dietary allowances of the countries in the world that recommend them. However, different countries have their own ideas on how much calcium is needed to maintain health in a healthy person which is why there is so much variation in the figures. When comparing recommendations, however, it is as well to bear the following points in mind.

Ranges are given in some countries (eg Australia, Caribbean, FAO/WHO, India, Singapore) but in the table the higher figures are given in each case. New Zealand quotes a minimum safe intake and an adequate daily intake. The latter, higher figure is given in each case. In addition, some countries subdivide the individuals in the different age groups in their recommendations into sedentary, moderately active, and very active depending upon their energy output. In all cases, however, the calcium intake suggested does not differ with the activity of their lifestyle. Some authorities also change their calcium recommendations in pregnancy with the three stages. In all cases, the figure for the third stage or (trimester) is quoted because this is always the highest.

The countries that do not give recommendations usually rely upon other's figures. Most African countries have no dietary recommendations but those that do tend to use FAO/WHO figures. Belgium has no national standards but the American, English, Dutch, or French tables can be used. There are no national standards in Greece but they use either WHO (1974) or USA (1974) recommendations. For Guyana, Caribbean standards are relevant. Iceland provisionally accepts USA suggestions but is expected to adopt the Scandinavian standards on a permanent basis.

Rumania has no national standards. In South Africa they used their own 1956 National Food Board recommendations, but they were never revised so now they accept the US and FAO/WHO standards. Sri Lanka uses the FAO/WHO (1974) figures. Surinam possesses no national standards but accepts those of the Caribbean. Sweden has recently made official the Scandinavian recommendations for herself. Switzerland generally uses USA standards with occasional slight deviations. In Yugoslavia the dietary recommendations are at present being revised.

The average amount of calcium actually eaten daily varies from 350 to 1,200mg with the quantity closely related to the milk supply. Views on the importance of dietary calcium have changed dramatically since 1920. This is apparent in the six editions of the textbook written by Professor H C Sherman of Columbia University between 1920 and 1950. He stated that 'the ordinary mixed diet of Americans and Europeans, at least amongst dwellers in the cities and towns, was more often deficient in calcium than in any other chemical element.' The professor had a great influence on the teaching of nutrition at the time, with the result that Americans of all ages increased their milk intakes markedly.

Sherman's conclusion was based on two experimental findings. First, rat studies indicated that calcium intake was a limiting factor in their growth. In the second, adult men and women often went into negative calcium balance when put on diets containing between 500 and 700mg of the mineral. Such conclusions however were queried when it was found that in countries with a poor milk supply which led to a low calcium intake human bones calcified normally with no increased liability to fracture.

The beneficial effect of milk on the growth of undernourished children, well established in the 1920s, was thought to be due to its calcium content. More recent research indicated that this attribute is related to the protein rather than the mineral content of milk. Confirmation that children could grow perfectly well on very low intakes of calcium (about 200mg daily) came from studies by Dr L Nicholls and A Nimalasuriya reported in the *Journal of Nutrition* (1939). Growing children in Ceylon (now Sri Lanka) were capable of maintaining a

Table 5 Recommended dietary intakes or daily amounts of calcium around the world in milligrams

	Age range (yrs)	ARGENTINA (1976)	AUSTRALIA (1979)	BOLIVIA (1968)	BULGARIA (1980)	CANADA (1975)	CARIBBEAN (1976)	CHILE (1978)	CHINA (1981)	COLUMBIA (1975)	CZECHOSLOVAKIA (1981)	DENMARK (1985)	FAO/WHO (1974)	FINLAND (1980)	FRANCE (1981)
Infants	0–0.5	500	700	400	600	500	400	600	400	400	700	500	500	–	–
	0.5–1.0	500	800	550	600	500	500	500	600	550	900	500	600	–	–
Children	1–3	750	800	450	900	500	500	500	600	450	900	500	500	500	600
	4–6	1000	800	450	1000	500	500	500	800	450	900	1000	500	500	700
	7–10	1000	1100	450	1100	700	500	600	800	450	1100	1000	500	500	700
Males	11–14	1000	1400	650	1200	1200	700	700	1000	650	1200	1000	700	700	900
	15–18	1000	1300	650	1200	1200	600	600	1200	550	1200	1000	600	600	1000
	19–22	700	800	550	950	1000	500	500	600	550	800	1000	500	500	1000
	23–50	700	800	500	1100	800	500	500	600	500	800	1000	500	500	800
	51+	700	800	500	750	800	500	500	600	500	700	1000	500	500	800
Females	11–14	1000	1300	650	1200	1000	700	700	1200	650	1100	1000	700	700	900
	15–18	1000	1300	650	1200	700	600	700	1000	550	1200	1000	600	600	1000
	19–22	600	800	550	950	700	500	600	600	500	800	1000	500	500	1000
	23–50	600	800	500	1100	700	500	500	600	500	800	1000	500	500	800
	51+	600	800	500	850	700	500	500	600	500	700	1000	500	500	800
Pregnant		2000	1300	800	1400	1200	1000	1200	1500	800	1500	–	1200	1200	1000
Lactating		1500	1300	900	1200	1200	1000	1200	2000	900	2000	–	1200	1200	1200

Group	Age	PHILIPPINES (1977)	NEW ZEALAND (1983)	NETHERLANDS (1978)	MEXICO (1970)	MALAYSIA (1973)	KOREA (1980)	JAPAN (1979)	ITALY (1978)	INDONESIA (1980)	INDIA (1981)	CENTRAL AMERICA & PANAMA (1973)	HUNGARY (1978)	EAST GERMANY (1980)	WEST GERMANY (1975)
Infants	0–0.5	600	600	100mg kg/wt	600	550	360	400	500	–	600	550	–	600	500
	0.5–1.0	600	600	"	600	550	540	400	600	600	600	550	600	600	500
Children	1–3	500	600	800	500	450	600	400	500	500	500	450	500	600	600
	4–6	500	600	800	500	450	600	400	500	500	500	450	500	600	700
	7–10	500	700	800	500	450	1000	500	500	500	500	450	500	800	800
Males	11–14	700	800	1200	700	650	1000	800	700	700	700	650	500	800	1000
	15–18	600	800	1200	700	500	1000	700	700	600	600	550	500	800	900
	19–22	500	600	800	500	450	600	600	600	500	500	450	500	600	800
	23–50	500	600	800	500	450	600	600	500	500	500	450	500	600	800
	51+	500	600	800	500	450	600	600	500	500	500	450	500	600	800
Females	11–14	700	800	1200	700	650	1000	700	700	700	700	650	500	800	900
	15–18	600	800	1200	700	500	1000	700	700	600	600	550	500	800	800
	19–22	600	600	800	500	450	600	600	600	500	500	450	500	600	700
	23–50	500	600	800	500	450	600	600	500	500	500	450	500	600	700
	51+	500	600	800	500	450	600	600	500	500	500	450	500	600	700
Pregnant		1000	1200	1300	1000	1200	1000	1000	1200	600	1000	1200	1000	1000	1200
Lactating		1000	1200	1500	1000	1200	1000	1100	1200	600	1000	1100	1200	1100	1200

Group	Age range (yrs)	POLAND (1969)	PORTUGAL (1978)	SCANDINAVIA (1980)	SINGAPORE (1975)	SPAIN (1980)	TAIWAN (1980)	THAILAND (1970)	TURKEY (1975)	UNITED KINGDOM (1979)	URUGUAY (1977)	USA (1980)	USSR (1980)	VENEZUELA (1976)	WEST PACIFIC COUNTRIES (1972)	REPUBLIC OF IRELAND (1983)
Infants	0–0.5	–	360	360	600	500	400	500	–	600	–	360	–	550	500	540
	0.5–1.0	–	800	540	600	600	500	500	500	600	600	540	1000	550	500	540
Children	1–3	1000	800	600	500	650	500	400	500	600	800	800	1000	450	400	800
	4–6	1000	800	600	500	650	500	400	500	600	800	800	1000	450	400	800
	7–10	1000	1000	600	500	650	600	500	500	700	800	800	1200	450	400	800
Males	11–14	1200	1200	800	700	800	700	600	600	700	1200	1200	1500	650	600	1200
	15–18	1400	1200	800	600	850	800	700	700	600	1200	1200	1500	550	400	1200
	19–22	800	800	600	500	600	600	500	500	500	800	800	800	450	400	800
	23–50	800	800	600	500	600	600	500	500	500	800	800	800	450	400	800
	51+	800	800	600	500	600	600	500	500	500	800	800	800	450	400	800
Females	11–14	1200	1200	800	700	850	700	600	600	700	1200	1200	1500	650	600	1200
	15–18	1000	1200	800	600	600	700	500	600	600	1200	1200	1400	550	400	1200
	19–22	800	800	600	500	600	600	400	500	500	800	800	800	450	400	800
	23–50	800	800	600	500	600	600	400	500	500	800	800	800	450	400	800
	51+	800	800	600	500	700	600	400	500	500	800	800	800	450	400	800
Pregnant		1400	–	1000	1200	1300	1000	1000	1000	1200	1200	1200	1500	1200	600	1200
Lactating		2000	–	1000	1200	1400	1000	1200	1000	1200	1200	1200	1900	1200	600	1200

positive calcium balance on these low intakes of calcium.

————Adaptation to low calcium intakes————

The Sri Lankan children in the above study were obviously able to adapt to the low calcium levels of 200mg in their diets. Later studies by Dr D M Hegsted and his colleagues (*Journal of Nutrition* 1952) indicated that Peruvians on different levels of calcium intake excreted less calcium than would have been expected from reports in the North American and European literature. These people had been on low dietary calcium intakes all their lives. A comparable study was carried out on South African Bantus by Dr A R P Walker and U B Arvidsson in Johannesburg (*Metabolism* 1954) who measured calcium intakes and found them to be no more than 300mg daily. When the subjects died, measurement of the calcium in their bones indicated completely normal values, proving that calcification had been unaffected.

It is also possible for those on conventional high-calcium western diets to adapt to low intakes of the mineral. This was proved by Dr O J Malm in 1958 (*Scandinavian Journal of Clinical Laboratory Investigation*) who studied healthy Norwegian men between the ages of 20 to 60 years over a period of one year or more. The men were prisoners so their diets and calcium intakes and excretion could be accurately controlled and measured. Of the 26 subjects, 22 adapted successfully to the low calcium intake. Ten of these adapted very quickly and the other twelve did so but rather more slowly. The figures for one of the slow adapters are shown here. All figures are mg calcium per day.

Table 6

Days observed	Intake	Urine	Faeces	Balance	Total gain or loss
210	942	238	605	+100	+21.0g
210	436	200	323	−88	−18.5g
196	454	209	252	−7	−1.3g

This case illustrates that even in one accustomed to high calcium intakes, a reduction in intake leads the body to adapt to lower levels. The reduced amount in the faeces indicated that more of the mineral is being absorbed once intake drops. The body conserves more also, as the reduced urinary loss shows. Adaptation in this case has taken more than a year but a steady balance is eventually reached.

CHAPTER 3

The metabolism of calcium

In the human body, the skeleton contains more than 90 per cent of the total calcium. The bones themselves make up about 14 per cent weight which are about 9kg in a young adult male. Within this weight is 1.2kg of calcium, 1.8kg protein, 2.25kg water, 0.45kg fat and 3.3kg of other minerals. The protein present acts as a matrix upon which the insoluble calcium salts (mainly phosphate) form a crystalline lattice. Although these calcium salts (known as hydroxyapatite) make up the main inorganic component of bone, the other minerals such as magnesium, sodium, potassium, fluoride and carbonate all contribute to the maintenance of bone strength, possibly by forming cross links between the hydroxyapatite crystals.

Fluoride and magnesium are particularly important in this respect and this is why small amounts of each are essential in maintaining healthy bones and teeth. We have seen that fluoride has been used in prevention and treatment of osteoporosis and there is little doubt that fluoride helps in preventing dental caries. Recent studies by Dr Guy Abraham have led him to conclude that magnesium is essential in keeping bones healthy and lack of this mineral may also be a contributory factor in the onset of osteoporosis.

The skeleton is not a static tissue and each bone is continuously being replaced and reorganized. This dynamic state was demonstrated as early as the eighteenth century by that great anatomist John Hunter. He fed animals a red vegetable dye called madder which was taken up by newly formed bone. He then showed that the dye gradually left the bone indicating a two-way movement between the skeleton and the rest of the body.

Much more recently it was possible to introduce radioactive elements into bone, first with labelled phosphorus and then with labelled calcium. Radioactive labelling is a way of measuring uptake into bone of these minerals since the amount being introduced is known and its presence within the bone can be calculated from the amount of radioactivity detected. In this way the amounts of calcium deposited in and lost from the bone daily can be accurately measured.

The technique indicated that in adult man the calcium turnover is normally from 400–600mg daily which means that in a healthy individual this amount is lost and replaced every day. This quantity represents about 0.05 per cent of total body calcium so a complete change of mineral within the skeleton takes several years to accomplish. In osteoporosis, calcium is lost but is not replaced so it is easy to see how this sort of turnover produces an insidious weakening of the bone. At the other extreme are growing children whose turnover rate of calcium is so much faster that their skeletal mineral may be completely replaced in only one year.

──────Bone growth throughout life──────

It is possible to assess the size of the skeleton at different ages and so the amount of calcium laid down is readily calculated. In early infancy about 150mg of the mineral is laid down daily but by the age of 3 years it drops to about 75mg before it starts its steady climb. The largest amount of calcium is deposited in bone during the adolescent growth spurt at puberty. In the study demonstrating this (I Leitch and F C Aitken, 1959, *Nutrition Abstracts and Reviews)* English council school girls were found to have a maximum accretion rate of 340mg daily, at 14 years of age. English council school boys reach their maximum rate at 16 years of age when they lay down 400mg calcium per day. These quantities stay in the bone so they are over and above the amount of calcium moving in and out of the bones. After this period of maximum accretion the rate decreases in both sexes to the low adult figures at age 20 years.

This study was on healthy well-fed children but where protein and energy intakes are low, the rate of accretion of calcium into bone is reduced because the protein matrix, called collagen, upon which calcium salts are deposited is deficient. Similarly low rates result when the supply of calcium is reduced by impaired absorption from the diet, for example in rickets due to vitamin D deficiency.

When children suffering from protein-energy malnutrition are given calcium supplements alone there is little improvement in the rate at which calcium is laid down in their bones and growth is slow. Once they receive protein, however, deposition of bone calcium assumes normal rates and growth continues. This study confirms observations in many parts of the world where poor diets prevent the normal growth of children that it is lack of protein and energy that is the limiting factor in preventing formation of the cellular protein matrix. Once protein is laid down it is calcified in a normal manner.

─────────Pregnancy and lactation─────────

When a woman is pregnant, the demands of the foetus mean that she has to supply it with up to 30 grams of calcium. Once her child is born, she will continue to lose 300mg calcium daily in her breast milk. All of this extra calcium should be supplied in the mother's diet but if not, she will draw upon the calcium reserves in her skeleton. Hence repeated pregnancies with long periods of breast-feeding in a woman whose diet contains little or no rich calcium sources (such as milk and dairy products) will cause her to lose more and more calcium from her bones leading eventually to demineralization and softening. This condition is called osteomalacia and should not be confused with osteoporosis which is a completely different problem.

In the lactating mother a daily loss of 350mg calcium will, over a normal period of breast-feeding, result in a gross loss of 100g of the mineral. Even if this were not made up from the diet, it is a small amount to lose from the vast reserves of the skeleton. Nevertheless there is evidence that compensation for these additional losses occurs to some extent with enhanced calcium absorption during the latter stage of pregnancy and during lactation. This is why there is usually no X-ray evidence of loss of calcium salts from the bones during lactation although, as we have seen above, multiple pregnancies combined with low dietary intakes of calcium will eventually take their toll and osteomalacia may ultimately result.

──The role of calcium outside the skeleton──

Although only about 1 per cent of total body calcium resides outside the skeleton, it has major roles in the functioning of nerves and muscles. In the skeleton, calcium, not unnaturally, is present as a very insoluble salt called hydroxyapatite, but in body fluids and cells it occurs as soluble calcium ions that are readily transportable. These calcium ions are positively-charged calcium atoms and this is why they are soluble. Whilst calcium occurs both inside body cells and outside them in the fluid that bathes them, there is a higher concentration outside the cells. It is this imbalance that enables calcium to function in the transmission of nerve impulses.

Calcium ions in and outside the cells and in the cell membrane itself are attached to specific proteins. When a nerve impulse arrives at the junction where nerves and muscles meet, it causes calcium bound to the protein to be liberated as free calcium ions. These ions then act upon the muscle fibres causing them to contract. When these calcium ions are deficient, as for example in rickets, nervous control of muscle is lost resulting in the condition known as tetany. Tetany (not to be confused with tetanus) is spasm and twitching of the muscles, particularly those of the face, hands, and feet. Treatment is simply supplementation with oral calcium, although the mineral may be injected intravenously for immediate relief of the acute condition.

Calcium ions in the blood are essential in the processes of blood clotting because they activate specific proteins that coagulate to form the blood clot. When blood is withdrawn from the body, the simplest way to prevent clotting is to remove the calcium ions by binding them to chelating agents such as citrate to produce insoluble complexes. Once the calcium is insoluble it cannot activate the clotting factors for transfusion or whole blood tests.

Calcium ions are also needed in the absorption of vitamin B12 from the diet. Small amounts activate the combining of the vitamin to gastric intrinsic factor and then are necessary for the complex to be absorbed into the intestinal cells. Usually there are enough calcium ions in the food itself or hanging around the intestinal mucosa for them to function in this way. There are no cases on record of anyone being unable to absorb vitamin B12 because of lack of calcium. The usual reason is a deficiency of intrinsic factor.

————Calcium in the blood————

Most of the calcium of the blood resides in the plasma, the clear, pale-yellow liquid in which the red blood cells are contained. It is very important for calcium to be maintained at a constant level in the blood plasma, and its concentration normally stays between 8.5 and 10.5 with an average of 10mg per 100ml. Within the plasma, calcium falls into three distinct compartments:

(i)   about 40 per cent of the total is bound to plasma proteins, primarily albumin.

(ii) about 10 per cent is combined with phosphoric and citric acids as calcium phosphate and calcium citrate respectively.

(iii) the remaining 50 per cent of the calcium is the easily diffusible ionic form.

It is this 50 per cent that exerts the physiological effects referred to previously. We shall see later how the effects of too high or too low levels of this calcium can produce problems with body functions.

Diffusible ionic calcium is difficult to measure in the blood plasma so if you have a blood calcium assay done this will be a total value. One complicating factor is that total plasma calcium is related to the levels of the blood protein called albumin and when this is low, so is the calcium. Hence if the blood albumin is reduced, as in some kidney and liver diseases, a low calcium figure is also seen. Despite this, the ionic calcium is normally maintained at a normal level so symptoms of low blood calcium are not seen. Hence the interpretation of the significance of any given value of blood plasma calcium is impossible without knowledge of the coincident concentration of plasma proteins. As a general rule, if blood plasma albumin deviates by 1 gram per 100ml from the normal value of 4.0 to 4.4 grams per 100ml, total blood calcium can be expected to change by 0.8mg per 100ml.

————Calcium exchanges in the body————

Calcium is absorbed mainly in the small bowel and usually about 30 per cent of that eaten is assimilated. Intestinal absorption involves the soluble ionized form of the ingested calcium and reflects at least two separate steps. In the first the mineral is taken into the cells lining the intestine and in the second calcium is transported across the cell then out into the bloodstream for transportation around the body. During its transportation calcium is attached to a specific protein. The production of this specific protein is under the influence of vitamin D so this vitamin is absolutely essential for the absorption of calcium from the intestine.

Before any attachment to this specific protein, calcium must be in the soluble form. Of course, it may already be in a soluble form but if not, the acid produced in the stomach will help dissolve any insoluble mineral present in the diet. Unfortunately, there are plenty of other factors present in the food and in the intestine, that have the opposite effect — they render the calcium insoluble by binding or reacting with it. These factors will be discussed later. The absorbed mineral joins the general miscible pool of calcium that is present in the blood and the fluids bathing body cells to the extent of 4 to 7 grams. In the lactating female 250mg daily is lost to the breast milk. In everyone, as much as 400mg per day is lost to the gastrointestinal tract in saliva, bile, and pancreatic and intestinal secretions. This calcium plus that which is unabsorbed from the diet makes up about 800mg per day that is lost to the faeces. Further losses occur regularly in the urine.

The urine normally contains between 100 and 350mg calcium per day. The amount is variable over a wide range from person to person and tends to be higher in the summer months. In women the level of excretion overnight increases after the menopause and it is believed that this may be one factor in the development of postmenopausal osteoporosis. A late night milk drink (rich in calcium) may help reduce this nocturnal bone calcium loss in these women. However, although milk is also rich in protein, too much of this nutrient may actually increase the excretion of calcium. A study by Dr J W Chu and colleagues reported in the *American Journal of Clinical Nutrition* (1975) indicated that urinary excretion of calcium falls when dietary protein is reduced and rises when it is increased. A cup of milk at night is beneficial without overdoing it. We shall see later how other non-dietary factors may lead to excessive losses of calcium via the urine.

Sweat contains only small amounts of calcium (15mg per day) and usually these are insignificant. However those working in extreme heat can lose over 100mg calcium per hour in the sweat according to studies reported in the *Journal of Nutrition* (Dr C F Consolazio et al 1962). Under such conditions the sweat can contribute about 30 per cent of the total calcium losses.

The miscible pool of calcium also provides for the slow turnover of the mineral in the skeleton. As we have seen, up to 700mg per day every day is passing in and out of the bones. All losses from the pool should be replenished by dietary calcium. The size of the pool is controlled by sensitive regulation of the concentration of plasma calcium.

——Regulation of blood plasma calcium——

The blood plasma calcium level is controlled within very fine limits by specific hormones that operate in the bones, the kidney, and the intestine. These hormones are three in number, namely, parathyroid hormone, thyroid calcitonin, and active forms of vitamin D. Although they each have separate effects upon calcium absorption, excretion, and deposition in bone, which is how they maintain plasma levels of the mineral, their interaction contributes to the fine control of these levels. The actions of calcitonin are generally opposite to those of the parathyroid hormone, sometimes called parathormone.

The parathyroid glands are two pairs of yellowish brown bodies that are situated in the neck, behind or sometimes embedded within the thyroid gland. When plasma calcium levels fall, this has an effect upon the parathyroid glands which respond by producing more of the parathyroid hormone. In its turn, parathyroid hormone causes a reduction in the amount of calcium being lost through the kidneys and at the same time it increases calcium absorption from the small intestine. Although these may in part be direct actions of parathyroid hormone it is believed that the main functioning is by the action of the hormone on vitamin D metabolism. Parathyroid hormone stimulates the conversion of inactive vitamin D to its active form and this causes the plasma calcium to rise, mainly by increasing the absorption of calcium from the intestine.

Once the concentration of calcium rises to a certain level, it has the effect of stimulating the production of the hormone calcitonin. This hormone is produced by the thyroid gland and it acts rapidly to lower plasma calcium by reducing the output of the mineral from the bone. In addition to this action though, there is antagonism between the two hormones and the increased secretion of calcitonin also causes a decrease in parathyroid hormone synthesis. As the concentration of parathyroid hormone decreases, blood levels of calcium will also decrease and the whole cycle starts again. These mechanisms are illustrated in the diagrams on pages 82 and 83.

In the healthy individual hormonal control of blood calcium levels carries on smoothly by a balance between the two mechanisms outlined above but there are cases where this control goes wrong and a low or high blood calcium level persists. These conditions are known as hypocalcaemia and hypercalcaemia respectively.

Hypocalcaemia

The prominent signs and symptoms of hypocalcaemia include tetany and related phenomena such as 'pins and needles', tingling of the nerves where they meet the muscles, closure of the larynx due to sudden contraction of the muscles leading to noisy indrawing of breath, muscle cramps, and convulsions. The low blood calcium that gives rise to these effects can be due to a variety of causes:

(i) Deprivation of calcium and vitamin D
This combination of deficiencies is observed in various conditions where absoption of nutrients from the intestine is reduced. It can also happen when the diet is inadequate in these micronutrients. If, however, there is malabsorption, then other micronutrients like

phosphorus and magnesium will also be at low levels in the body as well as the nutrients like protein. We have seen that low blood calcium stimulates the release of parathyroid hormone which in turn causes calcium to be released from the bone. When this happens in infants, they do not have sufficient bone reserves of calcium present to allow much withdrawal of the mineral before demineralization takes place. The combined loss of calcium and phosphorus therefore weakens the bones and the end result is rickets.

(ii) Hypoparathyroidism

Sometimes the parathyroid glands have reduced function or cease to produce parathyroid hormone altogether as a consequence of some genetic disorder or because of thyroid or other neck surgery. Obviously if the hormone is missing, then blood calcium levels fall, the bones have no stimulus to release calcium into the blood so the levels remain low. Sometimes a similar condition is due to a deficiency in production of the active vitamin D metabolite, but the end result is the same because calcium absorption from the intestines is reduced.

(iii)

Babies born to mothers who suffer from hypoparathyroidism often develop a temporary type of hypocalcaemia that gives rise to tetany. Sometimes this is the only clue to the mother's disorder. Other situations when the condition occurs in the newborn include high blood sodium levels and acute infections in the mother or babe. Once the baby is born however, its own parathyroid glands develop to the stage where it can maintain its own production of the hormone so the condition is often transient.

(iv)

Those with kidney disease can develop hypocalcaemia because their complaint increases the blood levels of phosphate which in turn decreases blood calcium levels. This is because the high blood phosphate level inhibits the conversion of vitamin D to its active form so the ability to absorb calcium from the diet is decreased.

The treatment of hypocalcaemia is normally supplementation of the diet with calcium salts at high concentration, or in severe cases the mineral — in the form of certain salts — can be injected directly into the muscle or blood. Injectable calcium is confined to calcium gluceptate, calcium gluconate and calcium laevulinate. They may also be taken orally in addition to calcium chloride, calcium lactate, calcium carbonate, calcium glubionate and dibasic calcium phosphate.

What happens when blood calcium rises

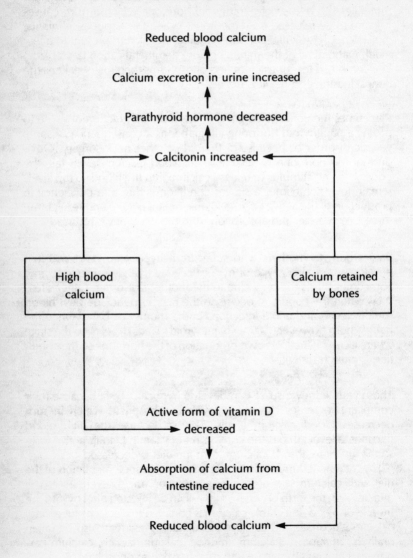

What happens when blood calcium falls

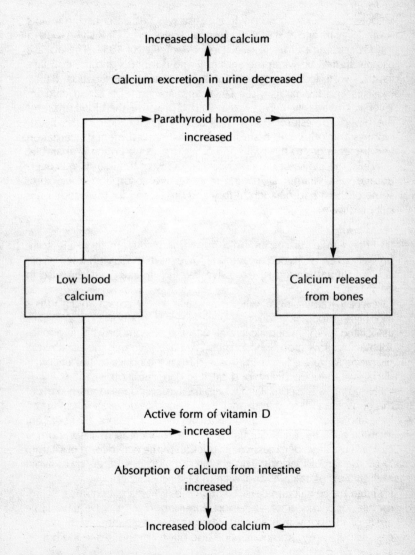

Hypocalcaemia

(i) *In infants*
Hypercalcaemia with no obvious cause was first described in infants some 35 years ago at St Mary's Hospital, London, by Dr R Lightwood in the *Proceeding of the Royal Society (Medicine)* of 1952. The disease usually started between the ages of 5 and 8 months and the constant features were loss of appetite, vomiting, wasting, constipation, flabby muscles and a characteristic facial appearance. The blood calcium level was invariably high as was the blood pressure, the blood urea and the blood cholesterol. There was usually calcification in the heart, kidneys and other soft tissues. There was marked mental retardation and the changes to the brain and other organs was often irreparable.

Whilst the cause of the complaint in many of the infants was overdosage with vitamin D leading to excessive absorption of calcium, some of them had no such history and the reason for the condition is still unknown.

(ii) *In adults:*
(a) High intakes of calcium in a normal individual in themselves are unlikely to cause hypercalcaemia. However in those with low thyroid activity (hypothyroidism) excessive calcium intakes can give rise to persistent high blood levels.

(b) Hyperparathyroidism, where the activity of the parathyroid glands is too high, producing too much parathyroid hormone, is classically associated with hypercalcaemia and the accompanying hypophosphataemia (low blood phosphate). We have seen that the parathyroid hormone stimulated calcium released from the bones so this becomes excessive, inducing high blood calcium concentration.

(c) Sarcoidosis is a complaint where the lymph nodes in many parts of the body are enlarged and small fleshy nodules develop in the lungs, liver and spleen. The condition is associated with about a 20 per cent incidence of hypercalcaemia. It is due to increased absorption of calcium because of excessive production of the active form of vitamin D by the sarcoid lesions. This is why sarcoidosis sufferers must avoid high intakes of calcium and vitamin D supplements.

(d) High blood calcium persists sometimes when a patient must lie immobile for a long time. It is most common in those who have large areas of the body injured and when extensive splinting with plaster casts is necessary. Immobilization also often contributes to the hypercalcaemia of patients with cancer.

(e) Hypercalcaemia has been reported in those with gastric and duodenal ulcers who also have impaired kidney function. The reason is that such people may be treated with a milk diet and antacids. The combination of milk and alkali, both of which are rich sources of calcium, is sufficient to increase the blood calcium level.

No matter what the cause is, hypercalcaemia is a serious condition because the mineral can be deposited in the kidneys with reduction of kidney function depending upon what is the reason for the condition. Usually agents that augment the excretion of calcium, such as certain diuretics, corticosteroids, calcitonin, and sodium phosphate are used as treatment.

Absorption of calcium — the role of vitamin D

No matter what form the dietary or supplementary calcium is in, its absorption from the intestine is completely dependent upon the presence of vitamin D. Therefore a discussion of the role of this vitamin is essential in any review of calcium absorption.

The metabolism of vitamin D

Vitamin D is a fat-soluble vitamin so, like any fat, it is absorbed from the food in the small intestine. The presence of bile salts is essential for its absorption because they are needed to reduce it to the tiny globules called chylomicrons. These are absorbed intact and as such the vitamin is carried to the liver attached to specific proteins in the blood. These same proteins carry the vitamin D that is formed in the skin (cholecalciferol) from the tissue into the bloodstream and hence to the liver.

For many years the vitamin was believed to act as cholecalciferol or as ergocalciferol but we now know that these are not the active forms. In the liver, both types are converted to the appropriate 25-hydroxy vitamin D (25 OH-D) (ie. either D_2 or D_3) and this is the form of the vitamins that circulate in the blood. Usually the concentration in the blood plasma is above 5 micrograms per litre. The fatty tissues of the body represent a greater storage depot for vitamin D than does the liver. It is excreted in the bile as inactive metabolites.

The 25-OH-D is carried in the blood to the kidney where a further conversion takes place under the influence of enzymes to the 1,25-dihydroxy vitamin D (1,25-$(OH)_2$-D). This is the active form of vitamin D and it is produced only in the kidney. When the kidney is not functioning correctly, as in renal diseases, this transformation is inefficient and it explains why one feature of such diseases is often an upset in vitamin D function.

The active form 1,25-(OH)$_2$-D (either D$_2$ or D$_3$) is some ten times more potent than the vitamin itself in its action on target tissues, and it acts much more quickly. In many studies it was noticed that there was a 10 to 12 hours delay before vitamin D exerted its functions and it was this observation that stimulated the search for a more active form. The goal was reached in Cambridge by Dr E Kodicek in 1974 and in Wisconsin, USA by Dr H F Deluca in 1976. The discovery of 1,25-(OH)$_2$-D opened up entire new frontiers in our knowledge of how vitamin D functions.

─────────The functions of vitamin D─────────

We know that vitamin D does not act as such but must first be converted to 1,25-dihydroxy vitamin D and this compound has a hormone rather than a vitamin function. A hormone is defined as a substance synthesized in one part of the body that acts upon another part, the so-called target organs. Cholecalciferol is produced in the skin then converted by the liver and kidneys into the active 1,25-dihydroxy cholecalciferol which functions elsewhere in the body. Hence, cholecalciferol satisfies the criteria for a hormone. Since it can be synthesized in the body and need not necessarily be supplied in the diet, cholecalciferol does not satisfy one of the criteria for a vitamin. This intriguing suggestion was put forward by Dr W G Loomis in an article in *Science* (1967). It bridges the gap neatly between vitamin and hormone function and in this respect cholecalciferol is unique amongst the vitamins.

Vitamin D, as 1,25-dihydroxy vitamin D, functions as a hormone which with two other hormones known as parathyroid hormone and calcitonin, regulates calcium and phosphate metabolism. Three metabolic effects of the active form of the vitamin have been identified.

1. It promotes the absorption of the mineral calcium in the small intestine, by inducing the synthesis of a specific protein required to bind the calcium within the cells lining the intestine.
2. It acts on bone, causing it to release calcium into the blood circulatory system where it can then be transported to where it is needed elsewhere in the body. This mechanism also requires the presence of parathyroid hormone.

3. It facilitates the absorption of phosphate from the small intestine by a similar mechanism to that operating for calcium. This is independent of the calcium transport system, however, since it operates in a different part of the small intestine.

Whenever there is a fall in blood plasma concentration of calcium, synthesis of parathyroid hormone by the parathyroid gland is stimulated. In its turn, parathyroid hormone causes the transformation of 25-hydroxy-D into 1,25-dihydroxy-D within the kidney. Once this active form of vitamin D is released from the kidney it acts upon the intestine to absorb calcium from the food and upon the bone to release the mineral and so plasma calcium increases. When the concentration of calcium rises to a certain level it has the effect of stimulating the production of the hormone calcitonin.

This in turn causes a decrease in parathyroid hormone synthesis. The net result is decreased assimilation of calcium, so the blood level of the mineral stays constant. As calcium is used its concentration in the blood gradually drops, and when it reaches a certain level the whole cycle of events starts again. A similar regulatory mechanism controls blood plasma inorganic phosphate concentration.

——Deficiency of vitamin D and its causes——

Deficiency of vitamin D during childhood leads to the development of rickets. Rickets is a word derived from the Anglo-Saxon word 'wrikken', which means to twist. The disease is characterized by a softening and deformity of the bones due to a lack of uptake of the mineral calcium. Calcium may be present in the diet but, because vitamin D is deficient, the mineral cannot be absorbed from the small intestine.

In adults, lack of the vitamin causes the disease known as osteomalacia, which means softening of the bones. The features that differentiate rickets from osteomalacia are due to the fact that in children, the ends of their bones are in a state of active growth. Hence, the bones continue to grow but they are not hardened by the deposition of calcium phosphate, and under the weight of the child they twist and bow to give the characteristic appearance of rickets. In adults, growth of the bones has ceased, so in osteomalacia the problem is mainly one of lack of absorption of calcium from the diet. In an attempt to maintain blood calcium levels, the mineral is withdrawn from the bones which become soft, weak, and painful.

Although the disease was known for many years, it was not until the mid-seventeenth century that the spread of rickets reached alarming proportions. Its increase coincided with the development of industrial cities at that time. Chimney smoke and high tenement buildings combined to cut out the sunlight and, as we now know, this removed one of the prime sources of vitamin D. Before 1900, the incidence of rickets in children of the poorer classes was as high as 75 per cent in the larger cities of the world, because lack of sunlight was not compensated for by increased dietary intake of vitamin D. Milk and dairy foods were beyond the reach of the poorer families, and as we have seen, there is no dietary source of vitamin D beyond those, apart from the more expensive liver and meats. The main item of these people's diets was cereals, and these exacerbated the lack of sunshine and dietary vitamin D by supplying large amounts of phytic acid. This acid immobilized the calcium in the diet so lack of the mineral was yet another factor in the aetiology of rickets in those times.

Despite the discovery of vitamin D in 1918, it was not until 1923 that Dr Harrietta Chick and her colleagues, working in Vienna (which was a blackspot for rickets), demonstrated unequivocally that lack of sunshine and lack of the vitamin in the diet were both factors in the development of rickets. Thanks to these studies, the incidence of rickets in the West declined dramatically between the two world wars. This was due to an awareness of cod liver oil as an excellent supplement for children, coupled with increased exposure to the sun due to a programme of slum clearance and a reduction in industrial and domestic chimney smoke. Sunbathing became popular, in part due to a public awareness of the value of the 'sunshine vitamin', as vitamin D came to be called.

Rickets today

The prevalence of rickets has been studied extensively in Glasgow because, even today, its incidence in that city is the highest in the UK. Only ten or so cases are seen each year but surveys of more than 4,000 Scottish infants carried out by Dr G C Arneil (1975) indicated that up to 7 per cent of children aged 1 to 3 years have minimal rickets. Most reach school age without any bone deformities, so the effect of the observed levels of vitamin D in the blood is not certain. Other big cities in the UK have slightly lower incidences of rickets than Glasgow.

There are in Britain as many as 25 per cent of children whose dietary intake of vitamin D has been calculated as less than 100iu daily. This is far below the recommended minimum of 400iu. The low intakes are reflected in the decreased blood levels of 25-hydroxy cholecalciferol which are usually at the danger level of less than 5mcg/litre. Despite this, these children show no clinical signs of rickets, and X-rays of their bones appear normal. Nevertheless, they are regarded as suffering from hypovitaminosis D (low body levels) and represent a high risk of developing rickets if their vitamin intake remains low during the growing years. Supplementation with cod liver oil appears to be the best insurance against developing the disease.

During the 1950s the children of Asian immigrants to Britain began to show signs of nutritional rickets. In children of European origin this type of rickets is very rare beyond the age of 8, according to a *Practitioner* article by Dr G C Arneil in 1973, but among Asians adolescent children as well as the younger ones were appearing with rickets. This cannot be entirely due to their skin pigments, since the more heavily pigmented West Indian children are much less afflicted with the disease and, in fact, gave an incidence no higher than white children in the same areas. One survey, reported by Dr N Ruck in *Proceedings of the Nutrition Society* (1973), found that the average daily intake of vitamin D by Asian, West Indian, and European children was estimated to be only 60iu, 72iu and 64iu respectively; figures that show no real significant difference. Hence, there must be other reasons why Asian children have a greater prevalence to rickets when in Britain, and one clue lies in the vitamin D status of pregnant Asian mothers.

In one trial, reported in *Lancet* (1975), Asian women at the time of delivery of their babies had an average blood plasma concentration of 25-hydroxycholecalciferol of only 7.6mcg per litre, a figure dangerously near that diagnostic of hypovitaminosis D. The corresponding figure for European women was 18.3mcg per litre.

A later review of 3,327 deliveries in 1978 confirmed that blood plasma levels of 25-OH-D in Asian mothers were consistently low at the birth of their children. As many as 81 per cent of the mothers and 30 per cent of their babies had body levels consistent with osteomalacia in the adults and rickets in the babies. Other studies indicated that in general, adult Asians living in India had a much higher status of vitamin D than their counterparts living in Britain. It was concluded that the high frequency of rickets and osteomalacia among Asians in Britain can be explained by their low intake of vitamin D in the diet and inadequate solar exposure.

There are many more studies confirming the high incidence of rickets in growing children, and of osteomalacia in adult Asians, and much effort has gone into ways and means of combating these vitamin D deficiencies. The culmination of this concern was a Working Party under the chairmanship of Dr Elsie Widdowson, one of Britain's top nutritionists, who reported their conclusions in 1980 in the document 'Working Party on Fortification of Food with Vitamin D', Committee on Medical Aspects of Food Policy (*Rep, Health, Soc Subj No 19*). They advised against fortification of food with vitamin D as a means of tackling the problem of rickets and osteomalacia in Asians living in Britain. The argument was a familiar one. Previous experience had shown that 40 years after Britain had conquered rickets in its own population there was an outbreak of infantile hypercalcaemia (too much calcium in the blood) due to excess vitamin D intake. The path between efficacy and toxicity of vitamin D is narrower than was originally thought. Individuals vary so much in their consumption of foods amenable to vitamin D fortification (chapati flour was one possibility) that it was impossible to determine a safe level of constant fortification. The committee concluded that the safest courses were regular supplementation with the vitamin at the individual level and a programme of health education. It was also suggested that windows should be made of glass that would let in light at the wavelength necessary to allow formation of the vitamin in the skin, and that fluorescent lamps should be developed which emit a light that will also activate the vitamin in the skin.

Osteomalacia

Osteomalacia is the adult counterpart of rickets. It often affects women of child-bearing age who become depleted of calcium by repeated pregnancies. It was once very common in the Middle and Far East. The main factors included the purdah system (complete covering of the body and restricted life indoors), cold winters, and a poor diet with little or no milk and dairy products. In Europe the incidence of the disease paralleled that of rickets, and improved social and environmental conditions were factors in conquering both diseases.

Today osteomalacia is still an important disease in Asia, and it is not uncommon in elderly women in Scotland. It is still present in many northern countries where the winter sunlight is insufficient to allow adequate synthesis of vitamin D to make good a dietary deficiency. Particularly prone are old people who are housebound because of physical incapacity. Sometimes osteomalacia is complicated by osteoporosis, which is a honeycombing of the bone due to excessive loss of calcium. One study, reported in the *Lancet* (1974), found that 34 per cent of women over 50 years of age with fractures of the femur showed features of osteomalacia. This is relatively easy to treat with calcium and vitamin D, but other factors are concerned in the development of osteoporosis, which sometimes accompanies osteomalacia, and this is far more difficult to treat.

Occasional cases of osteomalacia in younger women have been reported from several larger industrial cities in Britain. Most of them are of Indian and Pakistani origins and the disease is present more frequently in pregnancy. Many studies have indicated the value of supplementation at this time, usually at daily doses of between 400 and 1000iu vitamin D daily. These doses also are sufficient to prevent deficiency of the vitamin in the new-born child.

Chronic diseases of the digestive system which prevent the absorption of fats are often the cause of osteomalacia. Vitamin D is fat-soluble and is best absorbed in the presence of fats, so it too is affected by these diseases. Osteomalacia also occurs in certain forms of kidney disease where the diseased organ is unable to conserve calcium and losses of the mineral are excessive. It is also possible that kidney failure results in the loss of ability of that organ to convert 25-hydroxy-D to 1,25-dihydroxy-D. The consequence is deficiency of the active form of the vitamin and decreased absorption of calcium.

Drugs used in treating epilepsy, known as anticonvulsant drugs, can induce rickets in children and osteomalacia in adults when they are on long-term therapy. Phenobarbitone and phenytoin are particularly active in this respect. The reason is that these compounds cause changes in the liver enzymes necessary to convert vitamin D to 25-hydroxy vitamin D. Instead of producing the required metabolite, inactive compounds are formed, so requirements for vitamin D are increased. Treatment is carried out by giving supplementary vitamin D or its metabolite 25-hydroxy vitamin D.

————Symptoms of vitamin D deficiency————

The infant with rickets has often received sufficient dietary energy and may appear to be well nourished. However, the child is usually restless, fretful, and pale with flabby and toneless muscles which cause the limbs to assume unnatural postures. Excessive sweating of the head is common, weakened abdominal muscles that have lost their tone give rise to distension, which is increased by intestinal fermentation of undigested carbohydrates. Gastro-intestinal upsets with diarrhoea are common. The infant or child is prone to respiratory infections. Development is delayed, leading to late development of teeth and there is failure to sit up, stand, crawl, and walk at the early stages. Bone changes represent the most specific symptoms of rickets, and these are characterized by thickening of the growing ends of the bones. As soon as the child can stand on its feet, deformities of the shafts of the leg bones develop, so that 'knock knees' or bow legs become an obvious feature. Rickets, as such, is not a fatal disease, but the untreated rachitic child is a weakling with an increased risk of infections, notably bronchopneumonia.

In osteomalacia, pain is usually present and ranges from a dull ache to the more severe variety. The areas usually affected are the ribs, the lower vertebrae of the spine, the pelvis, and the legs. Bone tenderness on pressure is common. There is often muscular weakness present, leading to difficulty in climbing stairs or simply getting out of a chair. Sometimes a waddling gait develops. Muscular spasms and facial twitching are the result of the low blood calcium levels that are a feature of the disease. The bones become brittle because of loss of calcium and spontaneous fractures are common.

Therapy with vitamin D

The only diseases with an established therapeutic response to vitamin D are rickets and osteomalacia but benefits in osteoporosis have also been claimed.

In treating rickets the usual dose is between 25 and 125µg (1,000 to 5,000iu) per day, depending upon the severity of the disease. Children can be given halibut liver oil in a very small dose of 1cc, because this oil contains 30 to 40 times the concentration of the vitamin of cod liver oil. In severe cases, the high dose of 125µg is usually met by using synthetic vitamin D (D_2). In third world countries where medical care is available only occasionally, treatment consists of a single massive dose of vitamin D, usually 3.75mg (150,000iu) every six months or so. Fortunately, side effects are rare, probably because the child is so depleted of the vitamin that the liver and fatty tissues can accept and store this amount safely. Calcium is usually given at the same time, but this is far more effective if taken in the diet on a daily basis.

The treatment of osteomalacia follows similar principles. When the disease is due to simple lack of dietary vitamin D (combined with reduced synthesis in the skin), a daily intake of 25 to 125µg (1,000 to 5,000iu) is sufficient to treat the complaint until symptoms disappear. Then a maintenance dose of 400iu daily is given, along with dietary advice and exposure to the sun, especially if the cause of osteomalacia cannot be determined or removed. When the patient is unable to absorb vitamin D, because of an inability to absorb fats, a much higher dose is required. A daily intake of up to 1.25mg (50,000 iu) then becomes necessary, and the vitamin must be injected into the muscle in order to bypass the intestine. When osteomalacia is related to kidney disorders even this dose is insufficient, and two or more times the amount may become necessary.

Osteoporosis (honey-combing of the bone) is the most common bone disease, and it particularly affects women past the menopause. It is due mainly to loss of bone calcium being faster than its replacement. Evidence has been presented by a team from Harvard Medical School, USA in *The New England Journal of Medicine* (1981) that one factor in the disease is a defective production of the hormone 1,25-dihydroxy vitamin D. This would explain why vitamin D itself is usually of little therapeutic use in osteoporosis but there is now hope that the active hormone 1,25-$(OH)_2$-D may be more beneficial.

Vitamin D status
─────────of Asians in Pakistan─────────
and the UK

We have seen that Asian immigrants in the UK are at risk of developing osteomalacia and rickets, although the actual number who develop these bone diseases is relatively small. Despite this, between 33 per cent and 44 per cent of Asian immigrants studied have biochemical evidence of vitamin D deficiency characterized by low blood concentration of 25-hydroxy vitamin D, the circulating form.

The specific reasons why Asian immigrants are more prone to vitamin D deficiency, when compared with the indigenous population of the UK, are not known with certainty. Various environmental and genetic factors have been implicated, notably inadequate exposure to the sun, dietary habits, skin colour, and possibly even some metabolic defect that reduced the efficiency of skin production of the vitamin. A study of the vitamin D status of Asians living in Pakistan compared to that of those living in Rochdale was thus undertaken by a group of medical researchers, and their results were published in the *British Medical Journal* (1983).

In a previous study of 262 Asians living in Rochdale, 90 had originally come from the Lahore and Rawalpindi areas of Pakistan. Enquiries in these areas led to the discovery of 19 first degree relatives of those who emigrated to Britain. The remaining 73 of the 92 people studied in Pakistan were recruited from neighbours of those relatives. They ranged in age from 11 to 75 years and were considered to be representative of, and comparable to, the Asians who had emigrated from that community. The serum 25-hydroxy vitamin D levels of all participants were measured. Clinical examinations were carried out to ascertain any overt signs of vitamin D deficiency, and in addition, a full dietary history was obtained for every individual in order to assess dietary vitamin D intake.

The control group of 92 Asians living in Rochdale were chosen such that their age and sex were identical to those in the study group. Their ages ranged from 11 to 60 years. Since 25-hydroxyvitamin D levels in the blood vary with the seasons, this group had their blood measured in April and a second group of controls, 62 in all, had their levels determined in the autumn. The results were quite enlightening.

None of the individuals living in Pakistan had clinical signs of vitamin D deficiency. Their blood levels of 25-hydroxyvitamin D were significantly higher than those in their UK counterparts, both in the spring and autumn groups. Women living in Pakistan had significantly lower blood serum levels of 25-hydroxy vitamin D than the men. There were no differences detected between city and village dwellers. However, calculated dietary intakes of vitamin D were lower, on average, in Pakistan than in Rochdale, despite the fact that the type of diet consumed was essentially the same in the two countries.

The higher serum 25-hydroxyvitamin D levels in those living in Pakistan indicates that their vitamin D status is generally excellent. It therefore looks as though adequate exposure to the sun by these people can maintain good blood levels of the vitamin, despite a small dietary intake. It is highly unlikely, on these figures, that Asians leaving Pakistan will enter the UK suffering from vitamin D deficiency. The findings thus confirm that the development of deficiency in immigrant Asians is the result of environmental change and not some inherent genetic trait of their race.

One clear conclusion to emerge from the study is that the principle source of vitamin D for Asians living in Pakistan is sunlight. There may, therefore, be several factors involved in producing lower blood serum levels of 25-hydroxyvitamin D in migrant Asians. One significant factor could be the reduced intensity of ultra violet light in Rochdale due to a greater cloud cover, a more northerly latitude, and perhaps a polluted atmosphere. Coupled with this, the Asians' preference for a secluded lifestyle and the observance of purdah (screening of women as part of their religious practices) may contribute to their decreased skin production of vitamin D leading to low serum levels of 25-hydroxy vitamin D. This practice would explain the lower levels of the vitamin in the women of Pakistan compared to their men. Even so, the higher intensity of ultraviolet light in that country would appear to suffice to keep the blood levels of these women at a reasonable concentration. All other things being equal, the switch to a UK climate would, not surprisingly, produce the lower blood serum levels of the vitamin.

The authors conclude that it is unlikely that education of the Asian community to increase their exposure to sunlight will be effective. There are moral objections to changing cultural practices that are related to religious beliefs. It is to be hoped that a switch to a western diet, with its increased intake of vitamin D, may serve to increase these people's vitamin status. Until these changes take place, however, the authors strongly recommend preventive measures against rickets and osteomalacia in the Asian community by the use of vitamin D supplements.

Although rickets is usually associated with a deficiency of vitamin D or its metabolites, there are cases on record where a lack of calcium in the diet has produced the complaint in an infant. Dr S W Kook and colleagues reported one such case in 1977 (*New England Journal of Medicine*). A 10 month old infant was receiving a diet based on lamb but devoid of all milk and milk-based products because of an intolerance to lactose. The diet provided only a calculated intake of 180mg calcium daily but there was no lack of vitamin D. Eventually the infant developed all the clinical, X-ray, and biochemical signs of rickets that responded only to the addition of an unfortified skimmed milk formula to the diet. Hence lack of dietary calcium, even in the presence of adequate vitamin D, can be a factor in the development of rickets.

The adult condition that is counterpart to rickets is osteomalacia and whilst this too is usually caused by a lack of vitamin D or its metabolites, there are cases where calcium deficiency is the culprit. Hence it happens in young pregnant women of Asian origin in Britain as it does in Asia itself where the cause has been identified as a combination of deficiency of vitamin D, because of low exposure to the sun that is exacerbated by a lack of the vitamin, and of calcium in the diet by avoidance of milk and milk-based products.

How does vitamin D function in controlling calcium absorption from the diet and in maintaining calcium uptake by the bones? Part of the answer lies in the presence of a specific protein, called calcium-binding protein or CaBP in the intestines, the kidneys, and the bones. These of course are the tissues that are very active in transporting calcium. One of the highest concentrations of CaBP occurs in the glands responsible for developing the eggshell in hens.

The importance of vitamin D lies in its ability to stimulate production of CaBP without which calcium cannot be transported from the intestines to the blood or from the blood to the skeleton. It probably also plays a part in carrying calcium from the blood through the kidneys. Hence if there is no vitamin D there is no CaBP and even in the presence of adequate calcium, this cannot be carried to where it is needed in the body. There is an interesting parallel here with the functioning of vitamin K, another fat-soluble vitamin. The sole function of vitamin K is to produce the proteins specific to the blood clotting process and in its absence the blood will not coagulate. Here then we have two examples of fat-soluble vitamins being responsible for the production of two different types of specific proteins.

————Other factors enhancing absorption————

Stomach acid

We have seen that calcium is absorbed across the intestinal wall as soluble ions which are then transported into the intestine cells attached to a specific protein. In the food, however, and in some supplements, calcium is presented in the form of relatively insoluble salts which must be solubilized before absorption. The agent for this is stomach acid which has always been regarded as a prerequisite for calcium absorption. The part of the intestine nearest to the stomach is the duodenum and it is here that the actual mechanism of absorption takes place.

What happens to calcium absorption when there is no acid in the stomach? This pertinent question arises because there are a number of conditions in which the stomach ceases to produce acid (a condition known as achlorhydria) such as pernicious anaemia, atrophic gastritis, and gastric cancer. Modern anti-ulcer drugs function by inhibiting gastric acid production so these too can cause the condition of achlorhydria, at least on a temporary base whilst treatment continues. In some people, however, achlorhydria is not associated with any disease and produces no ill-effects. How achlorhydria affects calcium absorption is therefore important in anyone suffering from the condition, whatever the reason.

Most of the studies on this problem have been carried out by Dr Robert R Recker of the Creighton University School of Medicine, Omaha, USA. His most recent report appeared in *New England Journal of Medicine* in 1985. The study group consisted of 11 patients with achlorhydria and 9 normal individuals. The patients' age range was 33 to 85 years with a mean of 62 years. Six of the patients had pernicious anaemia, and although their anaemia was being treated with vitamin B12 injections, their achlorhydria persisted. The other 5 patients had their achlorhydria confirmed by gastric juice aspiration. The control group were aged from 40 to 70 years with a mean age of 59 years. All participants had calcium intakes of about 800mg daily.

Two calcium preparations were tested. One was powdered calcium carbonate presented in hard gelatine capsules and providing 250mg of calcium plus a tiny amount of radioactive calcium. The second calcium salt under test was calcium citrate, providing 250mg of calcium, plus the same tiny amount of radioactive labelled calcium as in the calcium carbonate capsules. One capsule of each preparation was given at different times to all subjects and blood and urine samples were obtained 24 hours later. Measurement of the total calcium and the radioactively-labelled mineral in these enabled the amount of calcium absorbed in each case to be calculated. Experiments were carried out and the extent of absorption was measured in a fasting state and with a standard breakfast of eggs, toast, fruit juice, and coffee.

The results were highly significant. Whilst in the control group the calcium in both calcium carbonate and calcium citrate was absorbed to a similar extent, there was a great difference in the achlorhydric patients. Based on the blood measurements, ten times as much calcium was absorbed from calcium citrate as from calcium carbonate. The absorption value for calcium from carbonate was significantly lower in the achlorhydric group than in the control group and was lower than the value for calcium from citrate in either group. Similarly, calcium from citrate absorption was significantly higher in the achlorhydric subjects than in the normal group as well as being higher than the value for carbonate in either group.

The absorption of calcium when it was taken with a meal was little affected in the control group but there were profound differences in the achlorhydric group. Absorption in these patients was much higher in the presence of food than in the fasting state, to such an extent that it approached that of normal subjects.

Poor absorption of calcium due to achlorhydria may affect a large proportion of individuals who are at risk of osteoporosis. It has been established that acid output decreases considerably after the age of 60 years (M I Grossman et al, *Gastroenterology*, 1963). Other studies (P M Christiansen, *Scandinavian Journal of Gastroenterology*, 1968) have shown that the incidence of achlorhydria increases with age. Grossman's study reported that over 10 per cent of women over the age of 59 have no gastric acid production even after stimulation by drugs. Nearly 40 per cent of these women were deficient in free acid in

the basal state. The higher figure is the more important one because other workers (P Ivanovich et al *Annals of Internal Medicine*, 1967) had previously shown that calcium was not absorbed from calcium carbonate when basal stomach-acid secretion was zero in their subjects. After gastric acid stimulation with drugs though calcium absorption did increase.

These results have great significance for the type of calcium salts used in supplementation therapy. This is widely recommended for menopausal and postmenopausal women as a preventative against bone loss of calcium and subsequent osteoporosis, and the most widely used supplement is insoluble calcium carbonate. Because of the high prevalence of achlorhydria in the menopausal population the uncertain absorption of calcium in the fasting state makes the carbonate less desirable perhaps than the soluble calcium citrate. However calcium carbonate is perfectly satisfactory in those with achlorhydria as long as it is taken with a meal. The acids naturally present in foods are sufficient to assist its absorption. This finding confirms the earlier studies of Dr G W Bo-Lim and his colleagues (*Journal of Clinical Investigation*, 1984) who reported that calcium absorption from calcium carbonate was similar whether the stomach contents were neutral or acidic as long as food was present.

Milk

According to *Dietary Intake Source Data*, United States, 1971–74, Hyattsville, Maryland, National Center for Health Statistics, 1977 (DHEW publication no PHS 77–1221) about 50 per cent of US women between the ages of 18 and 70 years ingest less than 500mg calcium per day. The recent National Institute of Health Conference on osteoporosis recommended a calcium intake of 1,000 to 1,500mg per day to reduce the incidence of osteoporosis in postmenopausal women. Where then is the extra calcium to come from? The usual recommendations are from increased intakes of milk and milk-based foods and from supplements.

We know that milk is a rich source of calcium, but how well absorbed is it? Various interested parties claim that calcium is absorbed better from milk products than from supplements because the lactose in milk stimulates the transport of calcium by the cells that line the intestine. This claim is based on many animal studies where lactose was shown to increase calcium absorption and on studies in normal human beings who exhibited similar results. The people who lack the enzyme lactase, ie those who cannot digest lactose, showed, in some studies, a reduced capacity to absorb calcium, presumably because the enzyme contributes in some way to the absorption process. It must be said, however, that other studies have failed to find a stimulating effect of either lactose or milk on calcium absorption.

A study from the Department of Internal Medicine, Baylor Medical Center, Dallas, USA, by Dr M S Sheikk and his colleague and reported in the *New England Journal of Medicine*, 1987, has compared calcium absorption from milk and various supplements. The subjects studied were eight healthy men (aged 25 to 30 years) none of whom were lactose-intolerant. Calcium absorption was measured by allowing all subjects to fast for eight hours before their entire gastrointestinal tract was cleansed throughout with a poorly absorbed solution. Four hours after the lavage was completed each subject was given a calcium salt, milk, or a non-calcium-containing placebo. All the calcium sources were swallowed with water. Four hours later the subjects ate a meal and six hours later a second lavage was performed to remove all the unabsorbed material from the gastrointestinal tract. The calcium content of this effluent was then measured. The amount of calcium absorbed was calculated from the following equation:

Nett calcium absorbed = Calcium eaten in supplement minus
(Calcium in effluent — calcium in effluent after placebo).

The meal given to the subjects underwent the whole procedure seven times: six times with various calcium sources and once with placebo.

The results indicated no significant difference among the five calcium salts and from milk. The actual figures of percentage calcium absorbed were 32 from calcium acetate; 32 from calcium lactate; 27 from calcium gluconate; 30 from calcium citrate; 39 from calcium carbonate, and 31 from whole milk. What turned out to be highly significant was that the extent of absorption bore no relationship to the solubility of the calcium salts. All the calcium from salts with widely differing solubilities in water was absorbed to the same degree. The most likely explanation was that the acid in the fasting stomach was sufficient to dissolve all of them to the same degree. Even when the stomach acid was partly neutralized by ingesting food, the resulting mildly acidic medium was able to solubilize the calcium salts completely in one hour.

Although the study was confined to young healthy males there is no evidence from other reports that the extent of absorption of calcium in older men and from females, young and old, is any different from the subjects in this study. Even though elderly persons may absorb a lower total amount of calcium, the relative calcium absorption from different sources is likely to be similar in normal old and young people. In this study at least, calcium from milk was no better absorbed than that from widely-used supplements, and even amongst these there were little differences in the extent of calcium absorption in individuals with normal secretions of stomach acid.

Protein

As long ago as 1920, Dr H V Sherman was studying the calcium requirements in men and reported in the *Journal of Biological Chemistry* that an all-meat diet in conjunction with low intake of calcium resulted in increased urinary loss of the mineral. The conclusion that this calciuria was a result of the excessive protein intake was confirmed by the later studies of Dr R A McCance and Elsie Widdowson (1942, *The Chemical Composition of Foods*) who found that supplementing a conventional diet (providing 45 to 70g protein per day) with extra purified proteins did indeed cause excessive calciuria.

Other studies had indicated that high protein intake enhanced the intestinal absorption of calcium (the latest by Dr H W Linkswiler and colleagues, 1974, *Transactions of the New York Academy of Sciences Series II*) and it was felt that the urinary loss of the mineral on such diets was more than offset by increased uptake from the diet. What appeared to be a satisfactory state of affairs was upset by later studies that indicated that there was no significant increase in calcium absorption in response to high protein feeding (e.g. Y Kim and H W Linkswiler, 1979, *Journal of Nutrition*). The reason for these discrepant observations apparently lay in the actual amount of protein eaten. When protein was added to a low protein diet then an increase in calcium absorption was generally noted. If extra protein was added to a diet already replete in protein there was no additional absorption of calcium. A more quantitative approach than before also indicated that any increase in calcium absorption associated with excessive protein in the diet was not sufficient to offset the increased calciuria.

How then does protein in the diet affect the calcium balance of those eating such diets, and is it related to calcium intake? Attempts to answer these questions have been made by a number of researchers, and a typical study is that of Drs H M Linkswiler, C L Joyce and C R Anand reported in 1974 in the *Transactions of the New York Academy of Sciences Series 11*, volume 36 pages 2,429–2,433. In this study young men were fed diets containing three levels of calcium, i.e. 500, 800 or 1,400mg per day, in combination with three levels of protein, namely 47, 95 or 142g per day for 15 days. The high protein levels were reached by adding a mixture of pure proteins to a low protein diet. In this way total protein intake was better controlled. The high protein diets were compensated by a high intake of phosphorus but magnesium intake was constant at 400mg daily.

The extent of calcium absorption when the mineral was given at the 500mg level was not affected by the amount of protein eaten at any level. However, more calcium was assimilated when 800 and 1,400mg of the mineral was eaten, but only with the medium (95g) and high (142g) intakes of protein; there was no additional increase in calcium absorption but the loss of the mineral in the urine did increase so all subjects then went into negative calcium balance.

How long does the increased loss of calcium in the urine last? Dr L H Allen and colleagues found that urinary calcium increased within 2 to 4 hours after a high protein meal (1979, *Journal of Nutrition*) indicating rapid absorption and excretion. This loss of calcium persists when high level (141g) of protein is eaten daily and excessive amounts in the urine continued over a 45-day period in male subjects according to Dr N E Johnson et al, 1970, *Journal of Nutrition*. Even with less heroic intakes of protein, calciuria was sustained over a 60 day period in a study of young women reported by Drs M Hepted and H M Linkswiler in the *Journal of Nutrition*, 1981.

The significance of the extent of dietary intakes of protein and its relation to calcium loss in the urine is far-reaching. By increasing protein intake from very low to very high amounts a rise of approximately 800 per cent in calcium excretion resulted, irrespective of calcium intake which varied from 100 to 2300mg per day (S Margen et al, 1974 *American Journal of Nutrition*). This means that an individual who eats high protein meals consistently can go into negative calcium balance to the extent of 137mg per day which is equal to a 4 per cent loss of skeletal calcium every year (L H Allen et al, 1979, *American Journal of Clinical Nutrition*).

The amount of protein eaten in this study was 225g daily and although few people habitually consume this quantity, the average adult consumption in the United States for example is about 111g per day. This means that a substantial portion of the population will consume more than this and a level of 142g per day is not unusual. As we have seen, this amount is sufficient to elevate calcium requirements beyond the usual intakes. High protein intakes must therefore be considered a risk factor for osteoporosis in the typical western diet.

Epidemiological evidence

There is little epidemiological evidence for an association between high protein diets and loss of bone calcium. It has been proposed by Drs A Wachman and D S Bernstein (*Lancet* 1968) that the metabolism of excessive intakes of protein produces a heavy acid load internally that causes dissolution of bone and subsequent osteoporosis. This idea does seem to be confirmed to some extent by studies on Alaskan and Canadian Eskimos. These communities consume a low calcium, high phosphorus, and high protein (200–400g daily) semi-carnivorous diet and they exhibit an unusually rapid rate of bone loss as they grow older (R B Majess and W Mather, *Human Biology*, 1975, amongst others).

Studies that have compared the bone mineral content of the meat-eating population (high protein) and that of vegetarians (lower protein and less acid-producing) could find no difference between the two, at least in females between 20 and 59 years. A similar study comparing male lacto-ovo-vegetarians (ie they eat dairy products and eggs but no meat, fowl, or fish) with their omnivorous counterparts indicated no difference in their bone mineral density at any age. When we look at females over the age of 59, however, in whom osteoporosis associated with age is more rapid than in males, there are distinct differences between omnivores (eat all types of food) and lacto-ovo-vegetarians. The latter had consistently higher bone mineral contents than the former, suggesting that ingestion of meat does tend to reduce bone calcium.

It is, however, significant that the rate of bone mineral loss is similar in the United States population and those of Guatemala and El Salvador where the protein content of the food supply is substantially lower (S M Garn et al, 1969, *Clinical Orthopaedics and Related Research*). However these differences are not confined to protein but also apply to calcium intakes. It is highly probable that the higher quantities of calcium consumed by Americans compensate for the increased losses of the mineral induced by the greater amounts of protein eaten.

What then can we conclude about the relationship between high protein intakes and calcium metabolism? There is no doubt that very high intakes of pure, isolated proteins have been shown to induce excessive calcium excretion in the urine, leading to a negative balance. This increased urinary calcium is due essentially to a decrease in the amount of calcium retained by the blood flowing through the kidney and this in turn is a direct result of increased body acid produced from the sulphur-containing amino acids in the protein eaten. These remarks apply only to purified protein in the diet since less calcium is lost if the protein is eaten as meat, fish, or fowl.

The difference relates to the high phosphorus content of these meats so this mineral is acting to help retain calcium better than in pure proteins where phosphorus content is significantly lower.

The moral is to try and obtain your protein needs from the diet since in this way phosphorus too is being eaten and both nutrients combine to retain blood calcium. Bodybuilders, weight lifters, and athletes, who consume great quantities of pure protein would be well advised to ensure concomitant increases in their phosphorus and calcium intakes also, otherwise they could be increasing their chances of excessive calcium losses in the urine leading to negative calcium balance and ultimately osteoporosis.

Dietary factors interfering with the absorption of calcium

We have seen that lack of vitamin D decreases the synthesis of the specific protein carrier of calcium within the intestinal cells and so absorption of the mineral is adversely affected. There are other factors, however, whose presence (rather than absence) in the diet can apparently inhibit the uptake of calcium by the intestinal cells, and these can now be discussed.

Calcium-phosphorus imbalance

It has been claimed for many years that a great excess of either calcium or phosphorus in the diet interferes with the absorption of both minerals and the increased excretion of the mineral of lower concentration. Based on this premise, a certain ratio between the two minerals has been considered desirable in the diet. The calcium:phosphorus ratio in the US diets has been estimated to be between 1:1.5 and 1:1.6 which is somewhat different from the desirable ratio set at 1:1. The imbalance observed probably reflects the phosphorus intake of modern diets mainly due to the meat, fish, and poultry constituents of the US and other western diets plus the high phosphorus content of popular beverages and widely-used food additives.

It must be said, however, that the evidence for the alleged detrimental effects of calcium:phosphorus imbalances comes from rat experiments showing that a high dietary ratio of the minerals caused rickets in these animals. When human studies were carried out, it was found that phosphorus intake (mainly in the form of phosphate) had little or no influence on calcium absorption within the range that these minerals normally occur in foodstuffs. The classic experiment was carried out on six Norwegian men who were given phosphate by mouth in sufficient quantity to provide an extra 250 to 1,000mg phosphorus per day (over and above that in their diets) for periods ranging from 4 to 8 weeks. Despite the bias towards a high phosphorus:calcium ratio, all the volunteers maintained their normal calcium balance during the term of the study (Dr O J Malm, 1953, *Scandinavian Journal of Clinical Laboratory Investigation*).

A later study by E M Widdowson et al (*Lancet*, 1963) looked at the effect of giving phosphate supplements to breast-fed babies on their absorption and excretion of calcium, strontium, magnesium, and phosphorus. In no way did these high intakes of phosphorus interfere with the absorption of calcium. Neither was calcium absorption nor its balance in the body affected by very high phosphate intakes in the diets of adults, according to a more recent trial reported by Dr H Spencer and colleagues in *Federation Proceedings*, 1975. This trial confirmed and extended Malm's original work carried out in 1953.

Modern nutritional thinking now disregards the calcium: phosphorus ratio as an important factor in determining the uptake of calcium from the diet of adults. There is a proviso for the new-born baby, however. Whilst breast-fed babies can be assumed to be receiving the ideal ratio of the minerals in their feeds, this may not be the case with those infants on artificial milks when an imbalance can affect calcium absorption (E M Widdowson, 1965, *Lancet*). For the first few weeks of life at least, it is important to maintain the dietary calcium:phosphorus ratio at that of human breast milk and this should be in the region of 1.5:1. Baby milk manufacturers should be aware of this and no doubt have adjusted their formulations accordingly.

Phytic acid

Phytic acid, also known as inositol-hexaphosphate, is a constituent of the outer hulls of many cereal grains and it was suspected for many years that it is the main factor that inhibits the absorption of calcium in those on high fibre diets. It was believed that phytic acid bound the mineral to make it unavailable for absorption. This idea arose during the First World War when Dr E Mellanby observed that diets rich in cereals produce rickets in puppy dogs, suggesting that these diets prevented calcium uptake from the food. In 1925, Mellanby extended his studies to look at the growth rate of these animals when bread was the main component of their diet. Those who ate the most bread grew fastest but at the same time developed the most severe rickets. It was thought therefore that bread had growth-promoting properties in the puppies that caused the bone protein matrix to develop faster than calcium could be laid down to harden it.

This theory was disproved when this animal model system was used to compare the rachitogenic (ie rickets-inducing) effects of various cereal flours. Wholewheat flour was found to be more rachitogenic than white flour and oatmeal was the worst of all. Sometime, years later, the same researcher reported the culprit in these flours was phytic acid.

It was not until 1942, however, that R A McCance and E M Widdowson carried out their classical studies on man to confirm that the puppy dog experiment results applied also to human beings. They carried out metabolic balance studies on five men and five women who were investigated continuously for 9 months. Their diets included wheat flour products providing between 40 and 50 per cent of the dietary energy. The effects of white flour with an extration value of 69 per cent (used in making white bread) were compared with those from brown flour of 92 per cent extration. Bread, cakes, pastries, and puddings were all made from the appropriate flour. The calcium balances are shown in Table 7.

Table 7 Calcium balances in white and brown bread (all figures in mg)

	White bread		Brown bread	
Subjects	*Intake*	*Balance*	*Intake*	*Balance*
Men (5)	580	-27	589	-112
Women (4)	426	-8	520	-57

These figures applied over a 3 weeks study. On white bread the subjects were virtually in calcium balance despite calcium intakes regarded as low by modern standards. Once they transferred to brown bread, however, all subjects went into negative calcium balance, but the extent of this varied from a loss of 10mg per day in one person to one of 190mg per day in another.

As phytic acid was considered to be the factor responsible for these differences in maintaining calcium balance, the next set of experiments looked at the effects of adding sodium phytate to white bread to see if it could reproduce the adverse effects of brown bread. The experimental subjects received an intake of phytate that corresponded to the phosphorus content of brown bread — a convenient way to measure the phytate present. The results of this study are shown in Table 8.

Calcium absorption was calculated from the difference in calcium eaten and that excreted in the faeces.

Table 8 The absorption of calcium and phosphorus from white bread, with and without added phytate (all figures in mg)

	White bread	White bread and phytate
Calcium eaten	492	521
Calcium absorbed	133	24
Phosphate eaten	1191	2114
Phosphate absorbed	822	1284

The results indicated quite clearly that added phytate reduced the amount of calcium absorbed. The important conclusion was that it was probably the phytate in brown bread that contributed to the reduced calcium absorption noted in the earlier study, since the lower the extraction rate of the flour, the less phytic acid (or phytate) there is in it.

The final proof that phytic acid reduces calcium absorption in this series of experiments (all reported in 1942) came with the reverse approach. What happens if brown bread is treated to remove its phytic acid content? Six subjects took part and the study lasted 3 weeks. The results are shown in Table 9. Again, the calcium absorbed was the difference between the amount of mineral eaten and that excreted in the faeces.

Table 9 The intake and absorption of calcium from different types of bread

Type of bread	Intake (mg/day)	Absorption (mg/day)	Absorption %
Brown	562	36	6.5
Dephytinized brown	610	165	27.3
Dephytinized brown & minerals	518	196	37.7
White	498	190	38.2

The removal of phytic acid also caused the removal of the minerals potassium, nitrogen, and magnesium as well as some calcium. In the third set of figures, these minerals were added back to the dephytinized brown flour before the bread was made. There is no doubt that the figures indicate that dephytinization (with or without the lost minerals) led to a marked improvement in calcium absorption. Hence the substantial increase in the calcium absorbed was regarded as confirmation that phytate was responsible for reducing calcium absorption. The other minerals appear to play no part.

Studies carried out some thirty years later indicated that when phytic acid is fed to subjects accustomed to a high-phytate diet, there is a reduced inhibiting effect on calcium absorption. (J G Reinhold et al, 1973, *Lancet*). This less-marked effect is believed to be due to adaptive changes that occur in the intestine that can induce synthesis of an enzyme called phytase which is capable of destroying phytic acid and so removing its inhibitory activity. This enzyme is known to occur in whole cereals but at differing levels. Oatmeal for example has a low phytase content and at the other end of the scale that of rye is particularly high. Mellanby's earlier experiments had indicated that oatmeal is more rachitogenic than rye flour, and although no one at the time was aware of the presence of this enzyme, it could account for his classic observations. Even in man, oatmeal when fed exclusively as the main part of the diet, can lead to significantly lower levels of calcium in the blood.

These results are based on experimental conditions and in these studies the switch to diets was carried out in a sudden manner. Such adverse effects of phytic acid are unlikely to occur in diets eaten over a long period because of the adaptive mechanism discussed above. According to FAO/WHO report in 1962 (*Calcium Requirements* WHO Technical Report Series No 230) there are many communities throughout the world whose habitual diet is based on a very high intake of whole or lightly milled cereals that have been calculated to contain enough phytic acid to precipitate all the calcium it contains. These people show no evidence of lack of calcium in their bodies due to their high phytic acid intakes so they must have the ability to break down phytic acid or to split the calcium from the mineral-phytic acid complex.

It is highly unlikely therefore that calcium absorption is influenced to any great extent by the amounts of phytic acid commonly encountered in modern diets. With excessive intakes of bran or raw cereals however, some effect may be seen but as we shall see later phytic acid is not the only inhibiting factor in these foodstuffs and dietary fibre may play a part. Nevertheless, in some Asian communities, the large amount of phytic acid present in chapattis — which are a major item in the diet, especially when they are made from wholemeal flour — can be a contributory factor in the diminishing absorption of calcium leading eventually to rickets and osteomalacia.

Bread is less inhibitory on calcium absorption than whole wheat grains and cooked cereals less than the raw variety. It is therefore likely that during the action of yeast on flour and the subsequent baking of the bread, phytic acid is destroyed and any phytic acid-calcium complex is split to release the calcium. At the same time, the combination of yeast and mild heat will stimulate the action of phytase. This would explain why calcium is more available for absorption from leavened than from unleavened breads.

Dietary fibre

The inhibition of calcium absorption associated with high intakes of whole grains has generally been believed due to their phytic acid content. However, it has also been observed that plants low in phytic acid but with high dietary fibre levels have an inhibitory effect on calcium absorption, suggesting that dietary fibre itself can bind the mineral. Dietary fibre is not a single entity but is a multi-component mixture of plant cell walls that include several types of polysaccharides that are not digested by normal body enzymes. The make-up of dietary fibre therefore varies from plant to plant and even with the growing conditions and age of any particular plant. Studies on the effect of dietary fibre-containing foods such as bran and the refined single fibres that are its constituents are discussed below.

Table 10 Dietary fibre is made up of the following constituents

Cellulose	This is made up of glucose units, as are starch and glycogen, but unlike these two polysaccharides, the glucose units in cellulose cannot be liberated by the digestive system. The cell walls of fruit and vegetables contain lots of cellulose.
Hemicelluloses	These are made up of sugars other than glucose, and like cellulose, hemicelluloses cannot be digested to their constituent sugar units. Fruit and vegetables are rich in hemicelluloses.

Lignins These are polymers of aromatic waxy alcohols that serve to toughen up the cellulose and hemicelluloses of plants as they thicken.

Pectins In the plant their function is to hold water in the inter-connecting network of cellulose in maintaining the structure of the plant. Pectins combine with water to act as thickening agents of which the best-known is apple pectin, used in jam making.

Gums These are water-soluble polysaccharides that dissolve in water to produce a thick, non-digested gel. They are known, along with pectin, as soluble dietary fibres. The best known are gum arabic, guar gum, carob gum, tragacanth and xanthan gum.

Mucilages These are polysaccharides found in seeds and in seaweed which is concentrated into dried kelp. Like the gums and pectin, they function because of their water-holding capacity and ability to produce gels that cannot be digested. Such substances are used as bulk laxatives and typical examples are ispaghula husks and psyllium seeds.

Many inconsistencies exist among the results of human studies on the effects of fibre on calcium bioavailability. In 1974, Drs K W Heaton and E W Pomare observed that wheat bran lowered blood plasma calcium (*Lancet*) suggesting that absorption of the mineral was adversely affected. The refined fibre cellulose added to bread resulted in a negative calcium balance according to Dr F I Beigi and colleagues reporting in *Journal of Nutrition* (1977). In a study on adolescent girls, the addition of cellulose to a low fibre diet increased the faecal excretion of calcium resulting in lowered blood serum levels of the mineral (R Godara et al, 1981, *American Journal of Clinical Nutrition*). Negative calcium balances were reported for bran-supplemented diets by J H Cummings et al (1976, *American Journal of Clinical Nutrition*).

In direct contrast to these studies, the addition of 20 grams of wheat bran (two-thirds of an ounce) to the daily diet of elderly people was found to have no effect on blood serum calcium (I Persson et al, 1976, *Journal of the American Geriatric Society*). Nor could Drs K W Heaton, A P Manning and M Hartog (1976, *British Journal of Nutrition*) find any adverse effects of wheat bran on calcium absorption in their young male subjects. Confirmation of lack of effect has come also from many other studies, of which that by Dr D J Farrell and colleagues (1978, *Australian Journal of Experimental Biological Medical Sciences*) is typical. He could find no significant differences in faecal calcium when a conventional diet was bran-supplemented.

When the high fibre diet is supplied by fruit and vegetables, faecal calcium is increased, whilst the urinary levels of the mineral equals that of a low fibre diet, suggesting that absorption of the mineral is affected adversely by such a diet (J L Kelsey et al, 1979, *American Journal of Clinical Nutrition*). However, no effect was noted in further studies on diets composed of fruit and vegetables by the same researchers (1981, *Journal of Agriculture and Food Chemistry*). Only when fibre was added to a high protein diet were the dietary calcium requirements found to be increased in his subjects by Dr H H Sandstead and his many colleagues reporting in *Dietary Fibers Chemistry and Nutrition* (Academic Press, New York, 1979).

No one is quite sure why these results are so contradictory. One reason could be variation in the dietary fibre constituents between the diets studied. The presence of high protein in the diet appears also to influence calcium absorption. Most importantly, the quantity of fibre could be a critical factor. For example, Dr Van Dokkum and his colleagues reporting in the *British Journal of Nutrition* (1982) could find no effect on calcium balance from feeding 22 grams of dietary fibre to his human subjects for 20 days. Once this was increased to 35 grams daily, all his subjects went into negative calcium balance. No effect was noted in another study when 16 grams (half an ounce) of wheat bran was eaten daily by healthy human volunteers (A Sandberg et al, 1982, *British Journal of Nutrition*) but it must be remembered that wheat bran is only 44 per cent dietary fibre so their intake was only 7 grams.

People can adapt to high fibre diets. Such diets provide large amounts of uronic acids, which are polysaccharides that can hold on to calcium, making it unavailable for absorption. Pectin, for example, is very rich in uronic acid but the bacteria that inhabit the bowel are capable of releasing the calcium. The mineral can then be absorbed from the colon and the calcium balance restored.

It is important to know how soluble fibres like sodium carboxymethycellulose (prepared from cellulose), carob gum, cellulose itself, and karaya gum affect calcium bioavailability because such gums are increasingly being used in high intakes to reduce high blood cholesterol levels and in the treatment of diabetes. An important study was that reported by Dr Kay M Behall and her colleagues in the *American Journal of Clinical Nutrition* (1987). Eleven men consumed a basal diet with and without added fibre (one of those mentioned above) at the level of 7.5g per 1,000 calories eaten daily for a 4 week period. Food, urine, and faecal levels of calcium were all measured. What emerged was that adding refined fibres to the basal diet did not significantly affect the calcium balance in any case. Hence it would appear that long-term feeding of soluble fibres will have no adverse effect upon calcium assimilation from the food.

Perhaps the most sound advice on how much bran to take for its beneficial effects without adversely affecting mineral balance is that of Dr K W Heaton in the *British Medical Journal* of 11 October 1983. He states that:

> There is no evidence that adding bran to a mixed Western-style diet induces deficiency of any mineral or vitamin. Experiments showing that bran induces negative balance of calcium, zinc, or iron have all been short term and probably in the long term adaptation occurs, with reduced urinary excretion. Blood serum concentrations of iron, calcium, and zinc were no different in 68 people who had been taking 7–32g per day for 6 to 48 months and in 43 people eating a similar Western-style diet but without the addition of bran.

A daily intake of bran of two tablespoonfuls is unlikely to affect the mineral uptake from a sensible, mixed diet. Remember too that bran itself is a very rich source of the essential minerals.

Oxalic acid

Oxalic acid is a natural constituent of some foods and it can combine with soluble calcium salts in the diet to form the relatively insoluble calcium oxalate so rendering the mineral unabsorbable. The foods that contain the highest concentrations of oxalic acid are spinach, rhubarb, beet tops, swiss chard, and cocoa. However, as the amounts of these consumed in the average UK and western diets are seldom sufficient to supply significant quantities of oxalic acid, their effect on calcium absorption can usually be discounted.

High saturated fat intakes

Excessive dietary intakes of fats, especially those that are saturated, may depress calcium absorption. This is because such fats are digested to saturated fatty acids which combine with soluble calcium in the food to form insoluble calcium soaps — a process known as saponification. The insoluble calcium soaps are then carried unchanged down the digestive tract to be excreted in the faeces. This is why patients with chronic intestinal disorders such as sprue and coeliac disease, who have large quantities of undigested fat in their faeces lose significant amounts of calcium leading almost inevitably to osteomalacia. In children afflicted in a similar manner the end-result is usually rickets.

Medicinal drugs

A wide selection of commonly used medicinal drugs can affect calcium balance in anybody by a variety of different mechanisms. Amongst these are prednisone, a corticosteroid widely used in inflammatory and degenerative complaints, that adversely affects calcium transportation; diphosphonates, used in the treatment of Paget's disease, that decrease the formation of 1,25 dihydroxycalciferol, the vitamin D metabolite that is needed for calcium absorption; the sedative glutethimide, which accelerates the breakdown of vitamin D. Other drugs that adversely affect vitamin D metabolism, and hence reduce calcium absorption, are phenobarbitone, phenytoin, and primidone, all of which are prescribed as anticonvulsants.

Osteomalacia can be a secondary condition induced by long-term therapy with magnesium and aluminium hydroxide antacids. They impair phosphate absorption by precipitating dietary phosphates and making them unavailable for absorption. The end result is that both phosphates and calcium are low in the blood plasma leading to osteomalacia. The chances of this happening in the elderly are often exacerbated by their reduced, age-related capacity to absorb calcium.

Other drugs can increase the risk of osteomalacia or osteoporosis by mechanisms other than malabsorption of calcium leading to its imbalance. They may accelerate excretion of the mineral, interfere with its functions in metabolism or simply immobilize it. Such drugs include the diuretics frusemide and ethacrynic acid; the laxative phenolphthalein; the anti-gout drug colchicine, and the antibiotic neomycin.

The only way to overcome the adverse effects of these drugs on calcium absorption or metabolism is to eat foods that are rich in the mineral or to take calcium supplements. Where the effect of the drug is an indirect one, as in the cases where vitamin D activity is interfered with, then the vitamin should also be taken. Suitable quantities on a daily basis are 500mg calcium and 250iu vitamin D (in addition to normal dietary intakes), which are beneficial without being harmful. These are supplementary potencies but similar extra quantities can be obtained if preferred by careful choosing of dietary items.

CHAPTER 4

Therapy
with calcium

We have seen that adequate calcium, either in the diet or taken as a supplement is a useful preventive factor in reducing the chances of developing osteoporosis and perhaps even in treating the condition or at least slowing down its insidious onset. There are, however, other clinical conditions where the mineral may have a therapeutic action and these include the lowering of high blood pressure; reducing the risk of cancer of the colon, and controlling blood cholesterol levels. It is important to assess the evidence for each of these potential uses for calcium, since if the claims are valid, simple mineral supplementation could be offered as an alternative to the potent drug therapy currently in use.

Calcium and blood pressure

Blood pressure is always measured in terms of millimetres (mm) of mercury and expressed as two figures. The first, higher one reflects the maximum pressure exerted by the left ventricle of the heart on the walls of the blood vessels of the circulation. This is called the systolic pressure. The lower figure is the blood pressure during the relaxation phase between heart beats. It is dependent upon the elasticity of the arterial walls and is known as the diastolic pressure. In healthy people, the systolic pressure usually lies between 100 and 140m of mercury and that of the diastolic pressure between 60 and 90mm of mercury. If the figures are above these, the condition is regarded as high blood pressure or hypertension.

Indications that dietary calcium may influence blood pressure came first from epidemiological studies on drinking water. An association between lower death rates from cardiovascular disease in those living in hard water areas compared to populations in soft water areas was first noted in the 1950s but later studies have confirmed the findings.

In 1957 Dr J Kobayashi of Japan reported that the ratio of sulphate to carbonate in river water (a measure of acidity) was associated with the incidence of stroke in his country. He concluded that lowering the acidity of the water with the alkaline calcium carbonate might reduce this incidence but the mechanism was not known. It must be remembered that in the early years there was no information regarding the cardiovascular effects of calcium so any biological relationship was unknown. Nevertheless, a few years later, Professor H A Schroeder of the United States related mortality rates from cardiovascular disease with the average water hardness within individual states in the USA and reported his conclusions in the *Journal of American Medical Association* (1960). There was a graded decrease in mortality from cardiovascular complaints as the average water hardness increased and this relationship was stronger than for overall mortality. There was a 24 per cent increased risk of mortality from heart and stroke diseases where the water was soft compared to that in areas of hard water with a mineral content of 200mg per litre.

Schroeder's study stimulated many others in the United States to carry out similar studies and in general they all verified his findings. A study carried out in 42 states by Dr R Masironi and reported in *Bulletin of WHO* (1970) could find no negative correlation between water hardness and all diseases of the heart and blood circulation but there was a significant negative correlation between hypertensive heart disease and water hardness. What emerged from this study was that within small geographic units inconsistent results are often observed simply because the populations studied are not big enough. Even on small-scale studies though, there was some association but of a lesser magnitude.

In the United Kingdom too there were found to be significant negative correlations of water hardness with deaths from all types of cardiovascular disease, coronary heart disease, stroke and hypertensive heart disease. Studies in England and Wales spanned from 1961 (J N Morris et al, *Lancet*) to 1969 (M J Gardner et al, *British Journal of Preventive Medicine*) but all indicated the same association. However, just like the Masironi study mentioned above, one carried out within the UK failed to find a significant correlation between water hardness and deaths from cardiovascular disease or stroke but did demonstrate a significant negative correlation between hard water and hypertensive heart disease (C J Roberts and S Lloyd, 1972, *Lancet*).

A review by Dr G W Comstock in the *American Journal of Epidemiology* (1979) on national studies from other European and

Latin American countries has shown that in these countries too, there was a relationship between hard water and hypertensive heart disease; in other words the relationship is between some constituent or constituents of the water and high blood pressure. What these constituents may be was the next consideration but the influence of other possible factors could not be ignored.

Although there are disparate results relating water hardness to cardiovascular diseases most of the studies indicate that a constituent of hard water, perhaps calcium, may in some way contribute a measure of protection from these complaints but in particular those related to high blood pressure. The following facts must therefore be considered:

(i)   even though beneficial effects on cardiovascular mortality may be a consequence of calcium and/or magnesium provided by the water, the amounts contributed are small compared to those found in the usual diet

(ii)   the hardness of water can be due to either calcium or magnesium or both, and which one contributed to it was not always determined

(iii)   the main risk factor for heart disease and stroke is hypertension, but only a few of the studies have correlated this condition to hard water

(iv)   it is difficult to measure the intake of calcium or magnesium from water by individuals since no account was taken of the use of home water softeners nor were individual water intakes measured.

These considerations have meant that investigators have switched their studies from hard water to specific dietary nutrients and their possible correlation with blood pressure. In this way one nutrient can be studied, its intake can be easily measured and any beneficial result can be related to that nutrient. This approach has been applied to calcium but most studies have been carried out only in the last five years.

It was in 1982 that Dr D A McCarran and his colleagues implemented a pilot dietary survey in a population of subjects, some with normal blood pressure and others with hypertension, living in Oregon, USA. They measured dietary calcium intakes by 24-hour recalls (this is where the subject reports everything eaten during the previous day) and found that calcium intake was 24 per cent lower in the 46 subjects suffering from hypertension (668 $\pm$ 55mg) than in the 44 subjects who had normal blood pressure (886 $\pm$ 89mg). In all, 19 nutrients in the diet were measured, but of these calcium and magnesium were the only ones to differ significantly between the blood pressure groups. The main difference for reduced calcium intake in the hypertensive subjects was lowered levels of non-liquid dairy products in their diets. At the same time more of the hypertensives had no liquid milk in their food intakes either. This study was reported in *Science* (1982).

A retrospective study on dietary data analysis of adult residents in California collected between 1972 and 1974 was carried out in 1983 by Dr S Ackley and co-workers and reported in the *American Journal of Clinical Nutrition* of that year. Milk and dairy product consumption were assessed independently and blood pressure measurements taken. The results indicated a difference between the sexes. In men calcium intake from milk was less in those with high blood pressure than in the ones with normal blood pressure values. Both systolic and diastolic pressures were inversely related to dairy product calcium intakes, i.e the higher these were, the lower the two pressures.

In women there was no relationship between blood pressure and dairy product consumption. Despite these differences, the authors concluded that some component of dairy products, presumably calcium, exerted a blood pressure-lowering effect in hypertensives.

These two preliminary studies stimulated a great deal of interest in dietary calcium and blood pressure and many other surveys were carried out both directly and in retrospect. The most comprehensive survey of lifestyle and health was The Health and Nutrition Examination Survey I (HANES I), carried out between 1971 and 1974 and conducted by the US National Center for Health Statistics. This included blood pressure measurements and an assessment of dietary intakes by 24 hour recall.

Various research groups analysed the data separately and the conclusions revealed were these:

(i) Across all age, race, and sex groups, calcium intake was the most consistent dietary difference between those with high blood pressure and those with normal blood pressure.

(ii) Low dietary calcium intake was associated with higher blood pressure, as were higher alcohol consumption and higher body weight. Dietary calcium eaten was significantly higher in persons with normal blood pressure than in those with high systolic pressure readings or with both systolic and diastolic hypertension.

(iii) Further analysis indicated that calcium intake was significantly related to systolic blood pressure only in non-white males but was not a significant predictor of systolic pressure in all groups. The primary predictors overall were age, race, and obesity. The only common mineral factor in all groups was the potassium levels; the lower the intake the higher the systolic blood pressure.

(iv)   At low calcium intakes, the strongest correlation to blood pressure was dietary sodium/potassium ratio and alcohol consumption. At high calcium intakes, alcohol was the only factor in determining high systolic or diastolic pressure.

Hence in this particular analysis, concurrent low calcium intakes are necessary for the sodium/potassium ratio to be an important dietary factor. It is therefore possible that calcium intake modifies the known relationship between the sodium/potassium ratio and blood pressure control.

A search of the literature indicates that there have been 17 epidemiological studies since 1983 confirming an inverse association between dietary calcium and blood pressure. The studies have come from all regions of the United States and some parts of Europe so a wide area has been covered with similar results. Significantly, this relationship has also been found in diverse ethnic groups, e.g Japanese males in Honolulu and Hawaii and Puerto Rican males in their own country, indicating a possible important role for calcium in controlling blood pressure.

Other national programmes that have studied the relationship between dietary calcium and blood pressure have confirmed the results from HANES I. In the Multiple Risk Factor Intervention Trial (MRFIT) carried out in the United States and reported by A W Caggiula et al (1986) at the 26th Conference on Cardiovascular Disease Epidemiology, American Heart Association, there was a significant relationship between calcium/energy ratio and both systolic and diastolic blood pressure. In the Honolulu Heart Program, milk was found to be the main source of both calcium and potassium, and calcium from all dairy sources was significantly associated with blood pressure. The Western Electric Study Data (M Nichaman and colleagues, 1984, *American Journal of Epidemiology*) revealed that when dietary calcium was high, blood pressure was low, but other factors that were related to the blood pressure included alcohol (negative) and polyunsaturated fats (positive).

One important problem with all these studies is the fact that calcium was not the only dietary factor related to blood pressure. The study by Dr H W Gruchon and colleagues reported in the *Journal of American Medical Association* (1985) was the only one where a single factor, dietary sodium, was consistently related to blood pressure. High sodium levels and low potassium intakes in the diet both appeared to be correlated with high blood pressure. Any connection between calcium and potassium may result from the fact that both are present in high amounts in dairy products and these foodstuffs can contribute 75 per cent of daily calcium and 40 per cent of daily potassium. It is important to note that there are at least three epidemiological studies indicating that calcium intake was no different between normotensive and hypertensive individuals.

Retrospective Studies

Evidence that dietary calcium does influence blood pressure has come from retrospective studies between 1971 and 1974 that were analysed in 1985 (M Yamamoto and L Kuller, *American Journal of Epidemiology and Circulation*). Low calcium intake by women in Pittsburgh, USA, was found to add to the risk of stroke. Since this condition is strongly related to hypertension, it can be assumed that any adverse effect of low calcium is manifested through increased blood pressure.

It is important to realize that no study has ever indicated a prospective link between dietary calcium intake and the risk of high blood pressure in the future — all associations have been shown in retrospective analysis of previously determined data. There is always the possibility that dietary changes have occurred in the years between nutrient measurements and development of a disease and these changes have gone unreported. In addition, blood pressure levels are known to change with time. Despite these objections, however, there is a certain consistency in the association between dietary calcium and blood pressure that has only previously been seen with alcohol intake and body weight.

It is not easy in epidemiological studies to discover a relationship between a single nutrient and a clinical disease because of a number of confounding and interfering factors. In any population under study there are bound to be variations in dietary intakes because of age, sex, race, and lifestyles. And of course it is impossible to assess any one nutrient in isolation because any diet will provide many nutrients and micronutrients. Let us therefore consider how these factors may relate to calcium intakes.

The most important variable in a population is age. In virtually all industrial societies the amount of calcium eaten in the diet decreases as age progresses. At the same time we know that blood pressure also increases as one grows older. Hence, if age is ignored in any study of calcium a false impression may be gained in any conclusions about the relationship between the mineral and blood pressure.

The sex of individuals is another important factor determining calcium intakes and blood pressure. It is well established that women in the west have significantly lower dietary calcium intakes than men but at the same time their prevalence to hypertension in the period of their lives up to the menopause is lower than in man. Once the menopause is reached, however, the rate of blood pressure increases with age in women is significantly greater than in men of similar age. Any study must therefore take gender into account.

Racial characteristics can also be confounding factors in any population study. Observations in the United States have shown that blacks consume significantly less calcium in their diets than whites, in both males and females. At the same time, blacks are known to have a higher prevalence of hypertension across all age and sex groups than whites. We must therefore add race to the other confounding factors, age and sex, as important considerations in assessing any relationship between certain intakes and blood pressure.

Lifestyles vary as do physical characteristics. For example, there is little doubt that body weight is important since obesity is one established causal factor in the development of hypertension. At the same time one study at least (D A McCarron et al, 1984, *Science*) has shown that dietary calcium is inversely related to body weight. This characteristic must therefore be considered and allowed for in interpreting epidemiological observations.

Other lifestyle variables need to be assessed in such studies but not necessarily adjusted for. Physical activity for example has been independently associated with lower blood pressure according to Dr R Cade and others reporting in *The American Journal of Medicine*, 1984. We have also seen previously that moderate exercise is a beneficial factor in promoting calcium balance in osteoporosis. The positive effect of exercise may therefore be extended to keeping the blood pressure down. Although alcohol intake has been linked with high blood pressure in many studies, whether there is any relationship between alcohol consumption and dietary calcium is not known. What has been established is that alcohol per se causes excessive calcium loss in the urine and tends to decrease bone mineral content by some mechanism unknown (H Spencer et al, 1978, *Osteoporosis II*, ed by US Barzel pub by Grune and Stratton, New York).

We must not forget either that other dietary nutrients may be of importance in the possible relationship of dietary calcium to blood pressure. We have seen that the amount of calcium absorbed is not directly proportional to the quantity eaten and this absorption is affected by other nutrients. Typical are phosphorus and caffeine which can adversely effect the assimilation of calcium according to a review by Dr L Allen in the *American Journal of Clinical Nutrition,* 1982. The same publication points out that high dietary sodium intake, mainly as common salt, is prevalent in many communities, and this can increase urinary calcium losses and exacerbate a low calcium intake.

Is the resulting high blood pressure due to excess sodium or deficient calcium? This is not known but at least one study in rats (D A McCarron et al, 1985, *Journal of Clinical Investigation*) found that a high level of calcium taken with a high level of sodium in their diets caused a greater reduction of induced hypertension than did high dietary calcium alone.

Environment

Two other variables are environmental in origin, and they are the actual calcium levels of the hard water being drunk and the amount of exposure to sunlight. Although most water authorities are aware of the mineral content of the water they control and supply, the analysis of these waters was not always taken into account during epidemiological studies. Anyone living in a hard water area may have their calcium intake increased by as much as 100mg per day from their water intake alone. Hence, if such calcium intakes were not taken into account during the studies, an underestimation of the mineral's intake could result. Hopefully, such underestimation applies equally to the whole population under study, including normotensives and hypertensives alike.

Exposure to sunlight is an important factor since, as we have seen, this induces the formation of vitamin D in the skin and this in turn is essential for the absorption of dietary calcium. Housebound individuals and others deprived of sunlight, and also those on low dietary intakes of vitamin D may be at risk of vitamin D deficiency and the resulting low calcium uptake could reflect in their blood pressure levels. This has not been shown by study but is a theoretical concept that in practice could confound any study on the relationship between calcium and blood pressure.

How significant could this relationship be in terms of the health of a community? The correlation found in the studies above was small in magnitude but nevertheless statistically significant. When dietary calcium intake was compared beween those with normal blood pressures and those with high levels, the difference in calcium intake was found to range from 5 to 25 per cent. These differences between highest and lowest calcium of dairy product intake resulted in a 4 to 6mm mercury systolic blood pressure drop. In a review paper by Dr G Rose (*British Medical Journal,* 1981), it was estimated that the modest reduction in blood pressure of as little as 5mm mercury on a population basis could result in a comparatively large decrease in mortality from coronary heart disease and heart failure. Hence if calcium can indeed induce even this relatively small reduction in blood pressure the consequences could be health benefits over the whole of the population.

————Calcium metabolism in hypertension————

If calcium intakes are related to blood pressure, we must assume that the mineral is controlling this through its own metablic actions. It is possible to measure biochemical parameters in blood, urine and bones in both normotensives and hypertensives and results to date indicate that there are differences. For example, the ionized form of blood serum calcium, although normally of low concentration, is decreased in those with high blood pressure compared to normal. Concomitant with this observation was the finding that dietary calcium too was lower in those with reduced blood levels (eg A F Folsom et al, 1986, *Hypertension*). We have seen previously that a low ionized calcium level in the blood stimulates the production of parathyroid hormone so this too should be increased in hypertensives. The fact that it is (eg P Strazullo et al, 1980, *Clinical Science*) confirms that the reduction in blood calcium is real in those suffering from high blood pressure. When calcium supplements were given to both normotensives and hypertensives, only in the latter were the blood ionized calcium levels increased (D A McCarron and C D Morris, 1985, *Annals of Internal Medicine*). At the end of 8 weeks, levels in the blood of both normotensive and hypertensive subjects were the same and this coincided with a small reduction of blood pressure in the hypertensives.

Another criterion for assessing the status of ionized calcium in the body is the level of this form of the mineral in saliva. When Dr H Maier and his colleagues measured this in hypertensive and normotensive subjects, the level in the hypertensives saliva was significantly lower than in the normals (1980, *Mineral Electrolyte Metabolism*). What is now beyond doubt is that even total calcium levels in the blood correlate positively with blood pressure, a finding confirmed in numerous studies.

Unfortunately, through some mechanism unknown, the excretion of calcium via the urine is higher in those with high blood pressure than in those with normal levels. Hence the low blood levels encountered in hypertensives are maintained by this loss and perhaps even exacerbated by it and a decreased dietary calcium intake. One might therefore expect that persistently higher urinary losses of calcium in hypertension would eventually lead to reduced bone density, but two groups of workers who studied this potential problem could find no difference in bone density between hypertensive and normotensive subjects (N E Johnson et al, 1985 *American Journal of Nutrition* and R Wasnich et al, 1983, *New England Journal of Medicine*).

Hypertension in pregnancy

Pre-eclampsia is the term applied to toxaemia of pregnancy, one characteristic of which is hypertension induced by the pregnancy. This too may be influenced by dietary calcium. The relationship between calcium intake and hypertension induced by pregnancy was first noted by Drs J M Belizan and J Villar (*American Journal of Clinical Nutrition*, 1980) who reported that populations with a high dietary calcium intake such as those of Guatemala and Ethiopia, have a low incidence of this condition. This calcium-pregnancy relationship does not appear to be related to the nutritional quality of the diet nor to the socio-economic conditions of those in these poor countries.

This epidemiological observation was followed up by the same workers (*American Journal of Obstetrics and Gynaecology*, 1983) who tested the effect of giving calcium to pregnant women. A total of 36 women with normal pregnancies were given either placebo, 1 gram of calcium, or 2 grams of calcium, daily during the second third (trimester) of their pregnancy. The results indicated that diastolic blood pressure levels were significantly less in the supplemented groups compared to the placebo one. Even throughout the last third of their pregnancy, the group taking 2 grams of calcium had significantly lower blood pressures than either the placebo or the 1 gram calcium per day groups.

These important findings were confirmed by a Japanese research group (N Kawasaki et al, 1985, *American Journal of Obstetrics and Gynaecology*) who obtained similar results but with only 156mg of calcium daily. They showed that not only was the incidence of pregnancy-induced hypertension reduced but the blood vessels of the treated women were less sensitive to high blood pressure-inducing drugs. One explanation is that the high levels of parathyroid hormones present during the last third of pregnancy, themselves capable of increasing the blood pressure, were reduced by calcium supplementation. Alternatively, the extra calcium may function simply by binding to cellular membranes and in this way calcium balance is maintained.

—Testing calcium supplements in hypertension—

The epidemiological studies quoted above do not prove cause and effect. They give a possible indication of a role for calcium in controlling blood pressure but the therapeutic benefit of the mineral can only be determined by intervention trials so that its effect or otherwise on high blood pressure can be determined. As is the usual with any new therapy, preliminary experiments are always carried out first on animals and these will be discussed as their results have a direct bearing on subsequent human studies.

Studies

(i) Normotensive rats

Strains of rats have been produced that are particularly prone to hypertension, either spontaneously or as a result of feeding, for example, excess dietary salt. These models therefore are extremely useful in studying any relationship between hypertension and calcium, especially as normotensive controls are readily available. An early report from Japan (Y Itokawa and colleagues, 1974, *Journal of Applied Physiology*) reported that when young rats were fed a calcium-deficient diet, they developed high blood pressure. This began to rise in the second week and it was 200mm mercury by the fourth week. The same rats exhibited cardiovascular effects and weight gain at the same time.

It is known that rapid growth renders young rats particularly susceptible to calcium lack and this of course may have contributed to the accelerated response. However, in studying adult pregnant female rats, Dr J M Belizan et al, reporting in the *American Journal of Obstetrics and Gynaecology* (1981), found that his normotensive rats also responded to a calcium-free diet with raised blood pressure. These blood pressures were readily reduced to normal simply by giving the animals adequate calcium in their diets. Even in strains of rats that developed hypertension on normal diets, extra calcium over and above their needs was sufficient to reduce their blood pressures to normal levels (A Berthot and A Gairard, 1980, *Clinical Science*).

The amount of calcium needed to reduce the blood pressure in these experimental rats is important. Even in animals with spontaneous high blood pressures, doubling the calcium intake will reduce the systolic blood pressure by 12mm mercury, (S Ayachi, 1979, *Metabolism*). In similar rats there was a graded blood pressure-reducing effect when calcium was added to their diets at three rates: 0.25 per cent, 0.5 per cent and 4 per cent. The more calcium given, the greater the reduction of high blood pressure. The effect also takes time. A period of treatment of only 2 weeks exerted no effect on the blood pressure.

We can conclude, therefore, that in experimental rats at least, calcium deprivation appears to increase the blood pressure of rats with normal values within a few weeks. Supplementing such animals with calcium causes a return of the blood pressure to normal but the effect is relatively slow, despite high levels of mineral supplementation.

(ii) Hypertensive rats

There are at least five ways of inducing high blood pressures in experimental rats and the result in each case has its counterpart in the human condition. The first is by feeding high salt diets — present evidence suggests that some human beings may also have salt-related hypertension. The second is by injecting a synthetic hormone called desoxycorticosterone acetate (DOCA), and again it is known that certain hormones can produce high blood pressure in man. The third is a combination of high salt intakes and DOCA. The fourth is by breeding genetically hypertensive rats from normal animals treated with hormone drugs — high blood pressure may also run in some human families because of genetic reasons. The fifth is called renal hypertension which is produced by partially obstructing the blood supply to the kidney of the animal. This is akin to the high blood pressure associated with kidney disease in human beings.

One significant finding to emerge from these studies was that DOCA hypertension can be prevented either wholly or partly by a calcium-rich diet as long as this is eaten before DOCA injection starts. When DOCA-treated animals on ordinary diets developed systolic pressures of 180mm mercury, their calcium-treated (double-intake) counterparts had figures of only 130mm mercury (J M Bertholt and A Gairard, 1980, *Clinical Science*). Confirmation of these findings came from studies where rats were given diets containing 2.8 per cent calcium carbonate (L M Resnick et al, 1986, *Clinical Research*) and in others where extra calcium (1.5 per cent calcium chloride) was presented in their drinking water (Y Kageyama et al, 1986, *11th Scientific Meeting of the International Society of Hypertension, Heidelberg*).

When rats were subjected to high salt diets to induce hypertension, the extent of the rise in blood pressure was reduced by incorporating 3.5 per cent of elemental calcium into their diets (P A Doris, 1985, *Clinical Experimental Hypertension, Theory and Practice*). This is a lot of calcium but it appeared to negate to some extent the blood-pressure raising effect of salt. There have been no such clear-cut responses when treating renal hypertensive rats with extra calcium since two studies from different research groups gave completely opposite results.

Other studies have indicated that rats with genetic hypertension that were fed calcium-rich diets developed the condition less rapidly and never reached the high blood pressure levels of their control counterparts who were fed stock diets. In quantitative terms, the rate of development of hypertension is inversely related to the amount of calcium fed — the higher the calcium, the more the hypotensive effect. Conversely, the higher the blood pressure the lower the calcium intake, even at levels that were just sufficient to maintain growth levels in these young rats (R Schleiffer et al, 1984, *Clinical Experimental Hypertension, Theory and Practice*). The most recent study on genetic hypertensive rats showed that weaning them onto a diet containing 2.5 per cent calcium as calcium carbonate reduced the final blood pressure at 20 weeks by 13mm mercury compared to the control group on a stock diet (F Perrot et al, 1985, *Archives Maladie Coeur*).

We can conclude therefore that in young rats with normal blood pressures, supplementation with oral calcium reduces these pressures by a small amount but in hypertensive rats this reduction is far greater. No matter which method is used to induce hypertension, there is a genuine reduction in blood pressure with calcium even when the condition is spontaneously-produced one. The only doubtful one is artificially-induced renal hypertension and this remains to be resolved one way or the other.

(iii) Human studies

Whilst animal model experiments can provide a lot of useful information on biological mechanism and even point the way to possible beneficial effects, it is only by controlled clinical trials in human beings that any therapeutic value of calcium in hypertension can be determined. High blood pressure is a symptom rather than a disease and its causes in human clinical conditions are many. The fact remains, however, that any treatment that can reduce the figures to normal values or prevent its development in the first place would be expected to improve the health of the individual being treated and perhaps slow down some disease progression. We shall now discuss clinical studies in human beings where hypertension is not induced artificially as in the rats but is a result of some dietary or metabolic influence.

The first human trial on calcium and blood pressure was carried out by Dr J M Belizan and his group and reported in the *Journal of the American Medical Association* (1983). The subjects chosen were 57 young, healthy students of both sexes and they were given either 1 gram of calcium (as the carbonate and gluconolactate) or a placebo daily for a total period of 22 weeks. Before the trial, all had normal blood pressure readings but at the end of it, those treated with calcium had lower diastolic pressures. The extent of the decrease was 5.6 per cent in the females and 9.0 per cent in the males. The females also exhibited a small though significant decrease in systolic pressure also. All of the blood pressures stabilized after about 10 weeks on calcium supplementation ie no further reduction occurred.

A second double-blind (ie neither subjects nor researchers know what is being given), placebo-controlled trial on 32 adults with normal blood pressures at the start, indicated that 1 gram of calcium (presented either as carbonate or citrate) daily caused a 3.3mm mercury decrease in diastolic pressure (D A McCarron and C D Morris, 1985, *Annals of Internal Medicine*).

Not all trials gave such promising results, and several research groups have reported a lack of effect of calcium on the blood pressures of people with normal values. One of these came from Holland, from the Netherlands Institute for Dairy Research and reported in the *American Journal of Clinical Nutrition*, 1986, by Van Beresteyn et al. The trial was double-blind and placebo controlled and carried out on 58 young, healthy normotensive females. Each received either placebo or 1,500mg calcium (as carbonate and citrate) daily for six weeks whilst on a low calcium diet that provided 500mg calcium daily. The results showed that in both the treated and placebo groups, blood pressure values decreased only slightly and no effect of oral calcium supplementation could be demonstrated. In addition, at the start of the experiment neither systolic nor diastolic blood pressure levels correlated with the subjects' habitual calcium intake. There was some correlation between diastolic, but not systolic, blood pressure and body weight at the start of the study but this was not affected by calcium supplementation.

The authors comment that the group on the low calcium intake (500mg per day) did not show any increase in blood pressure despite this dietary restriction. It is worth noting that the figure in this restricted group is also the Recommended Daily Amount of calcium for adults in the UK. The conclusion reached was that oral-calcium supplementation for 6 weeks does not influence blood pressure in young, healthy, normotensive females consuming a low-calcium diet (ie 500mg daily).

A much longer-term study was undertaken by Drs N E Johnson, E L Smith and J Freudenheim that was reported in the *American Journal of Clinical Nutrition* in 1985. The investigation was originally intended to determine the effect of 1.5 grams of elemental calcium (presented as calcium carbonate) on bone mineral content of 81 women. The supplement was taken daily by the subjects over a 4 year study period. Blood pressures were measured yearly but there was no evidence of any reduction in them due to the substantial added calcium. These women were all normotensive so it could be argued that calcium is more likely to be hypotensive only in those suffering high blood pressure.

Two more studies are worth a mention. In a hospital trial, Drs H Spencer and L Kramer were investigating calcium requirements and factors causing calcium loss. Despite increasing calcium intake in their patients from 0.2g to 2.5g per day they could detect no change in blood pressure (reported in *Federation Proceedings*, 1986). In one British study, an increased calcium intake from 0.2g to 1.8g per day failed to change the blood pressures of eight adults who had normal values (F P Cappuccio et al, 1986, *Clinical Science*).

The most recent report that indicates a lack of effect of calcium on the blood pressures of normotensive women comes from Margaret M Schramm and her colleagues from the Universities of Pittsburgh and Indiana, Indianapolis, that appeared in the *American Journal of Clinical Nutrition*, 1986. The subjects were 199 white women in the age range 46–66 years with no history of hypertension who were undergoing a trial assessing the effect of walking on postmenopausal bone loss. No calcium supplements were given but the mineral intakes of the women were calculated from a 3-day dietary study. Despite the fact that in the whole population of women calcium intakes varied from 300 to 3,000mg per day, with most of them in the 600 to 1,200mg range, there was no significant correlation between calcium intake and blood pressure detected. The authors concluded that dietary manipulation of calcium intake may not be beneficial in the prevention of treatment of hypertension in older women but they do admit that their patients were a select group of healthy women and may not be representative of the general population. It is also strange that their conclusion mentioned hypertension but all their subjects had perfectly normal blood pressures.

As with most research concerned with novel therapies, for every trial showing a positive benefit there is one that indicates a lack of activity and vice versa. Not surprisingly therefore there are many studies where calcium supplementation has caused a significant blood pressure decrease in those suffering from hypertension. The first, by Dr J M Belizan, has already been referred to above. The McCarron/Morris study mentioned previously looked at the effect of a daily intake of 1g elemental calcium (as carbonate and citrate) on patients with mild to moderate hypertension as well as normotensive controls. After 8 weeks supplementation, the systolic pressure was reduced by 3.8mm mercury in 48 hypertensive patients measured in the lying down (supine) position and by 5.6mm mercury in the standing position. Diastolic pressure (supine position only) was reduced by 2.3mm mercury. Similarly, in the N E Johnson study above, whilst normotensive patients were unaffected by calcium supplementation, those with high blood pressure had their systolic levels reduced by 20mm mercury, a highly significant figure.

An Italian study organized by Dr P Strazzullo and his colleagues and reported in *Clinical Science* in 1986 found a conclusive beneficial effect of calcium in reducing high blood pressure. Men and women with high blood pressure were given 1 gram per day of elemental calcium as a mixture of carbonate and gluconolactate for a total period of 15 weeks. This treatment significantly reduced systolic and diastolic pressures by 8.6mm mercury and 8.0mm mercury respectively.

Supplementation in older people with hypertension is even more effective in lowering the blood pressure. In American individuals with a mean age of 60 years, 12 weeks of treatment with 1 gram of calcium daily reduced the mean blood pressures by 4mm mercury. The systolic pressures in 30 per cent of the group were reduced by a highly significant 10mm mercury (C Morris and D A McCarron, 1986, *11th Scientific Meeting of the International Society of Hypertension, Heidelberg*). At the same meeting, Dr T R Ogihara and his team reported that in Japanese patients with a mean age of 73 years, 2 grams per day of calcium resulted in a drop of a massive 24mm mercury.

Other researchers have reported a more variable effect that appears to be dependent on differing calcium metabolism in their patients. On a daily supplement of 2 grams calcium, Dr L M Resivick et al (*Annals of Internal Medicine*, 1986) found that the response depended on the serum ionic calcium levels at the beginning of the treatment. If the calcium level was high, the blood pressure increased; if the serum calcium level was low, then blood pressure decreased.

The same phenomenon was observed in young people with mild high blood pressure by D E Grobbee and A Hofman from Erasmus University Medical School in the Netherlands. They reported in the *Lancet* (1986) on 90 young (16–29 years) mildly hypertensive patients who were given either placebo or 1 gram of calcium as calcium citrate daily for 12 weeks. There was no reduction in systolic blood pressure, but at 6 and 12 weeks diastolic blood pressure had fallen by 3.1 and 2.4mm mercury respectively more than in the placebo group. However those subjects with the higher parathyroid hormone levels (ie lower serum calcium) showed a greater fall in blood pressure to the extent of 6.1mm mercury diastolic after 6 weeks and 5.4mm mercury after 12 weeks than did the placebo group. In the words of the authors, 'The potential hypotensive action of calcium supplementation is of special importance in young people with mildly raised blood pressure in whom the pharmacological approach is unattractive.'

Those with hypertension who lose excessive amounts of calcium in the urine are more likely to benefit from calcium supplementation than people with lower rates of calcium loss. This was shown by Dr P Strazzullo and colleagues (*Clinical Science*, 1986) who studied hypertensive patients over a period of 15 weeks' supplementation.

We can conclude from these trials that oral calcium supplementation reduces blood pressure at least in a subsection of hypertensive people. This form of therapy was well tolerated, with few, if any, reported adverse effects. It is obvious that calcium has no short-term effect in those with normal blood pressure unlike its benefits in those with high blood pressures. The reductions are small in most cases but in some trials there were larger decreases. Such variable responses are well known in any therapy of hypertension. It is therefore worth a trial for anyone suffering from high blood pressure to take at least 1 gram of calcium daily for at least 12 weeks to observe if it has any beneficial effect. Such treatment can complement conventional drug therapy as some trials have shown. Research on human beings has been carried out only during the last five years so further research needs to be conducted. There are still questions to be answered about the long-term efficiency of calcium supplementation as a treatment for hypertension — about predicting those most likely to respond and the possibility of achieving a blood pressure reduction through sensible dietary approaches. It may even be proved eventually that adequate calcium intake during life can help prevent the development of high blood pressure. What is certain is that the recommended daily allowance of calcium may have to be reconsidered for the most effective prevention of two major problems of the elderly, namely osteoporosis and hypertension.

—————————Calcium and cancer—————————

Cancer of the colon and rectum are leading forms of the disease in western countries and epidemiological evidence accumulated over the past few years suggests that vitamin D and calcium are involved in protection against these conditions. One of the first epidemiological studies was carried out by Drs L Teppo and E Saxen of Sweden and reported in the *Israel Journal of Medical Science* in 1979.

These workers made a quantitative assessment of the incidence of colon cancer in the four Scandinavian countries Sweden, Norway, Denmark, and Finland. The incidences were expressed as new cases per 100,000 population per year from 1953 to 1959 and turned out to be:

Finland	8.0
Norway	12.9
Sweden	15.3
Denmark	17.7

These rates correlated inversely with the per capita daily milk intakes of 860, 554, 549, and 491 grams respectively. Whilst neither the incidence of cancer nor milk intakes were age-adjusted, it was reasonable to assume that colorectal cancers were most common in older men whereas milk consumption was higher in children and adolescents.

In Scandinavia, milk is not routinely fortified with vitamin D, and because all countries are in the northern latitudes, production of the vitamins by exposure of the skin to the sun is profoundly influenced by seasonal changes. The association of higher milk consumption with lower cancer incidence in Finland compared to Denmark has been confirmed by others (eg O M Jensen et al, 1982, *Nutrition in Cancer*). It must also be pointed out, however, that dietary fibre consumption in Finland is higher than that in Denmark and this food ingredient may be a complicating factor in comparing cancer incidence.

Studies in other countries have found a similar relationship between colorectal cancer incidence and the ingestion of vitamin D and/or calcium. Dr R L Phillips reported in *Cancer Research* (1975) on his studies on colon cancer incidence amongst Seventh-Day Adventists.

These people have a very low colorectal cancer incidence and a high milk consumption. In addition they live in Southern California where exposure to the sun is high and where milk is routinely fortified with vitamin D. Milk is a rich source of calcium and when it is presented in the diet with substantial amounts of vitamin D it is likely to be very efficiently absorbed. Hence it was believed that the combined calcium/vitamin D were the protective dietary factors against colorectal cancer.

Seventh-Day Adventists are not representative of the population of the USA because of their dietary practices, so other observations on the incidence of colorectal cancer in the country as a whole were carried out. The two Drs Garland (*Journal of Epidemiology*, 1980) who reported the survey, found that the death rates for colon cancer are greater throughout the nation in rural areas at the highest latitudes and in the major metropolises. The shorter days and indoor living and working lifestyles that characterize these respective regions suggest that lack of sun exposure could be a factor in the increased incidence of cancer. It was noted that deaths from bowel cancer for the period from 1959 to 1961 were inversely correlated with estimated daily solar radiation in metropolitan areas and at higher-latitude (ie northern) rural areas.

These observations led Dr Cedric Garland and his colleagues from the University of California and other US universities to carry out a prospective study of diet and health conducted at the Western Electric Company Hawthorne Works in Chicago from 1957 to 1979 in order to test the nutritional hypothesis that in the US calcium and vitamin D are protective factors against colorectal cancer. The subjects were 2,107 white men, mostly first or second generation immigrants from north-central Europe or longer-term residents of the USA of Anglo-Saxon ancestry. Incidentally, this was the same group of men who had been previously examined for a possible relationship between vitamin A and carotenes in cancer risk. The Garland study was reported in the *Lancet* (1985).

All subjects kept 28-day diet records at the time of initial medical examination and one year later. It was thus possible to measure energy, calcium, and vitamin D intakes in the diet, although no estimate of the vitamin from sun exposure could be made. Alcohol consumption and smoking histories were obtained and the macronutrients of their diets were assessed. Those who continued with the study were examined annually until 1969. A total of 1,954 men were included in the final analysis.

On the basis of medical records, medical examinations, and death certificates, four groups were established.

Group I   Those men with colorectal cancer
Group II   Those with other types of cancer
Group III   Those who died without any diagnosis of cancer
Group IV   Living men without known cancer

What emerged from the study was that the men in Group I had a significantly lower intake of calcium and vitamin D than those of the other three groups. These colorectal cancer sufferers also tended to be older and heavier than the men who were alive and free from cancer. Further analysis, taking into account smoking habits, age, and body weight, revealed a firm and significant inverse association between consumption of calcium and vitamin D and the incidence of bowel malignancy.

In the men with the highest intakes of vitamin D the risk of colorectal cancer was 16.4 per 1,000 men. Those with the lowest intake of the vitamin had an incidence of the disease of 30.7 per 1,000 men. When calcium intakes were compared with cancer incidence, the relative risks were even more significant. The men with the highest dietary intake of calcium had a colorectal cancer incidence of 12.3 per 1,000 compared to 38.9 per 1,000 in those with the lowest intake of the mineral. The conclusion was therefore that the dietary correlations in this study confirmed Garland's findings in the sunlight exposure survey and lend credibility to the idea that vitamin D/calcium is protective against colorectal cancer.

How the nutrients function is open to debate. The active form of vitamin D, 1,25 dihydroxy vitamin D, is known to have anti-cancer activity against human melanoma and leukaemia cells in tissue culture experiments. It is likely too that the enhancement of calcium absorption by vitamin D could play a part in its protective role. On the other hand, the main sources of both nutrients were milk and dairy products so it is possible that some other factor in high concentration in these foodstuffs could also contribute to their beneficial action. One such micronutrient is riboflavin (vitamin B2) but regrettably this was not measured in the study although no doubt it could be in retrospect.

Yogurt is another food that conveys special bacterial populations to the colon which are known to inhibit the effects of known colonic cancer-producing constituents (B R Goldin and S L Gorback, 1984, *American Journal of Clinical Nutrition*). Hence, if milk intake reflects yogurt consumption then the protective mechanism might be related primarily to bacteria rather than the constituents of milk. This, however, is purely speculative and must await much more research for confirmation. The simplest approach to reduce one's chances of getting colorectal cancer would appear from these studies to ensure good intakes of calcium and vitamin D from all sources throughout life.

These are epidemiological studies and the ultimate test for the proposed protective action of calcium and/or vitamin D against colorectal cancer must be a trial of calcium, for example, in those known to be at high risk of the disease. One such trial was carried out by Dr Martin Lipkin and Harold Newmark from the Memorial Sloane-Kettering Cancer Center, New York, USA and reported in the *New England Journal of Medicine* (1985).

It is known that in persons predisposed to colorectal cancer an early indication of abnormal activity of colonic epithelial cells is increased growth and spreading of these cells, and this can be detected by labelling such cells with a radioactive marker and determining their proliferation. Using the technique it has been established that significant differences in the frequency distribution of labelled cells exist between populations at high risk of familial colonic cancer (eg those where the disease is family-related) and others at low risk (eg Seventh-Day Adventist vegetarians). In the present trial, seven men and three women with a mean age of 56 ± 12 years who had a high risk of familial colonic cancer were studied.

The frequency and distribution of proliferating colon epithelial cells were measured in these individuals before and after oral supplementation of their diets with calcium carbonate (providing 1.25g elemental calcium). In the test period before calcium supplementation the profile of proliferating colon epithelial cells was comparable to that previously observed in subjects known to have had familial colonic cancer. However, two or three months after calcium supplementation was started, this proliferation was significantly reduced to that previously observed in people with a low risk of colonic cancer. It would then appear that oral calcium supplementation at the 1.25g per day level induces a slowing down of cell proliferation in subjects with a high risk of developing cancer of the colon. If this quiescent state can be prolonged for longer periods using simple supplementation, then perhaps the way is open to reduce the risk of the disease developing in those with familial colonic cancer. More studies are underway to determine the extent of this inhibition.

Calcium and cholesterol

We have seen that calcium supplementation may have a role in ameliorating hypertension, a well-defined risk factor for heart disease. Another risk factor for this condition is persistent high levels of fat in the blood usually associated with high blood cholesterol levels. Recent research suggests that here too calcium may be able to exert a protective function by reducing the level of high blood cholesterol.

High blood fats produce the condition known as hyperlipidaemia or hypertriglyceridaemia which is subdivided into a number of types (usually five) depending on the composition of the fats and the cause of the complaint. The most common is familial hyperlipidaemia which is due to a genetic discrepancy and tends to run in families. It is possible to induce hyperlipidaemia in experimental animals like rats and rabbits either by feeding high fat diets or by selectively breeding animals to produce a strain with the complaint. These animal models were used in the first experiments on the role of calcium in controlling blood fats.

Dr A Fleischman and colleagues (*Lipids*, 1972) in a series of experiments on rats fed high levels of fats showed that the serum cholesterol levels in these animals were reduced by 35 per cent simply by supplementing their diets with calcium. The total blood fats were reduced by 19 per cent with the same treatment. Other results of the mineral therapy were a reduction in body tissue fats and a decrease in faecal sterols and bile acids — all of which are produced from cholesterol. In view of the latter findings it is difficult to know where the lost cholesterol went to. In similar experiments with rabbits, a series of papers in the journal *Atherosclerosis* (1981 to 1983) reported that blood levels of cholesterol were reduced in these animals when they were fed calcium salts alone or a mixture of calcium and magnesium salts. At the same time, atheromatous lesions (ie fats deposited on the walls of blood vessels) were reduced in these animals, suggesting that even the deposition of fats can be reversed by calcium.

Translating these observations to human clinical studies also produced some promising results. For example, Dr H Yacowitz and colleagues reported in the *British Medical Journal* (1965) on 13 individuals who had high blood cholesterol and high blood triglyceride levels. When they took a supplement of calcium, providing 890mg of the element daily, for 21 days, their serum cholesterol decreased by 15.4mg per 100ml and their serum triglycerides by 32.2mg per 100ml. These reductions are significant and desirable but not always reproducible. Dr L A Carlson et al reported in *Atherosclerosis* (1971) that when their patients were fed 2 grams of calcium carbonate daily (corresponding to 800mg elemental calcium) for 8 weeks, their serum cholesterol was reduced by about 15mg per 100ml but there was no effect on serum triglycerides. The uneven responses may have been due to the two groups of patients suffering from different types of hyperlipidaemia. Those in the Carlson study had phenotypic hyperlipidaemia. Even after one year's supplementation, similar results were obtained.

The most recent clinical trial was by Dr Njeri Karanja and colleagues from Oregon Health Sciences University, Portland, USA, who studied changes in blood fat and cholesterol levels in their patients after calcium supplementation (*American Journal of Clinical Nutrition*, 1987). The study assessed the effect of the mineral on the blood fats in both normotensive and hypertensive subjects. Hence 43 patients with high blood pressure and 27 with normal blood pressures, all within the age range 21–70 years, received either 1 gram of calcium as carbonate or citrate or placebo daily for a period of 8 weeks. Half the subjects received calcium followed by placebo with the other half receiving the reversed treatment. Blood fats and nutrient intakes were measured repeatedly.

Overall, there were no significant changes in blood plasma fats as a result of calcium supplementation but this could be anticipated since most of the patients were relatively normolipidaemic (ie normal levels of blood fats). There were some benefits seen however when sub-groups results were studied. In those with normal blood pressures but with elevated serum cholesterol levels, supplemental calcium significantly decreased the levels from 250 ± 25 to 238 ± 20mg per 100ml — a drop of 4.8 per cent. What was of particular interest was that this reduction occurred in the LDL-cholesterol fraction since LDL can be regarded as the 'bad' cholesterol as opposed to the 'good' HDL cholesterol. Any decrease in LDL cholesterol can be regarded as highly desirable.

Subjects with mildly elevated levels of blood fats (lipids) and high blood pressure did not show any beneficial change in their blood fats in response to the calcium supplementation over the time the study was performed. However, an overall reduction of 3.8mm mercury (standing) and 5.6mm mercury (supine) systolic pressure was observed along with reduced supine diastolic pressure of 2.3mm mercury. Approximately 44 per cent of these same subjects responded to calcium supplementation at the rate of 1g per day with a 10mm mercury or more drop in systolic pressure. These findings would suggest that calcium's antihypertensive or hypotensive actions differ from that of its blood fat-lowering effect.

We can conclude therefore from this study that short-term supplementation with 1 gram elemental calcium can cause a significant reduction in the concentration of blood cholesterol in those with mildly elevated blood cholesterol levels and normal blood pressures. The extent of this reduction is comparable to that reported from the six-year Multiple Risk Factor Intervention Trial (*Journal of the American Medical Association*, 1982) which employed intensive dietary and lifestyle counselling coupled with drug treatment to reduce blood cholesterol and high blood pressure. It is therefore gratifying to see that simple supplementation with calcium can produce the same benefits.

What is less satisfactory is that blood fats and cholesterol are unchanged in those with high blood pressure accompanied by mildly elevated blood fat levels. Nor do the blood fats of those hypertensives with normal levels change with calcium supplementation although perhaps this is not so surprising. Why hypertensive, mildly hyperlipidaemic subjects do not respond is not known but response may be related to blood ionized-calcium levels. When these are near normal we know that blood pressures are reduced by extra calcium. Perhaps this takes precedence over any further beneficial actions of the mineral. If this is true then more prolonged supplementation or higher amounts of calcium may be of more use.

Children who have high blood fat levels do not appear to respond to calcium supplementation. In one trial (PHE Groot et al, 1980, *European Journal of Paediatrics*) 50 hyperlipidaemic children were supplemented with 1 gram of calcium daily but their blood fats were unaffected. There were some changes however when it was found that there were significant increases in their desirable HDL cholesterol fractions with a concomitant decrease in their undesirable LDL cholesterol fractions. Even this alteration in the spectrum of the different types of cholesterol may be regarded as a benefit in anyone suffering with hyperlipidaemia.

In another trial, Dr W D Mitchell and his team (*Journal of Atherosclerosis Research*, 1968) could not demonstrate any significant decrease in blood fats with additional calcium in his osteoporotic patients with normal blood fats. They did exhibit increased faecal levels of fats, cholesterol, and bile acids however and this response could be interpreted as beneficial since it did suggest that the body was ridding itself of excess fats and cholesterol. Perhaps simple supplementation with calcium is producing some immeasurable beneficial effect in those with high blood fats and cholesterol which would suggest that such individuals could help themselves by taking extra calcium daily.

————Osteoporosis — the final word————

In early October 1987, an International Symposium on Osteoporosis was held at Aalborg, Denmark, with an attendance of 1,300 delegates, reflecting the recognition that osteoporosis represents a major health problem in all developed countries. At the end of the symposium, a consensus development conference, sponsored by the European Foundation for Osteoporosis and Bone Disease, was held in order to formulate guidelines for the prevention and treatment of osteoporosis. A panel of 13 leading researchers listened to evidence from experts in public seminars attended by 700 people, including representatives of the medical profession, the pharmaceutical industry, the Press, and ministries of health. After deliberation by the 13 in a closed session, the panel discussed its report with the audience and a consensus statement was published in full in the *British Medical Journal* of 10 October 1987 and the salient points are presented below:

————Consensus statement————

Osteoporosis is a disorder characterized by a reduced amount of bony tissue per unit volume of bone. It is a major but not the sole cause of fractures in postmenopausal women and in the elderly.

Fractures occur most frequently at the wrist, the spine, and the hip with minimal trauma, and an osteoporotic woman may fracture any bone more easily than her non-osteoporotic counterpart.

Osteoporosis is an important health problem in the western world. More than 40 per cent of women will have experienced a fracture by the time they reach the age of 70. Moreover, the incidence of hip fracture appears to be increasing, and this is explained only in part by a longer life expectancy.

Osteoporosis results in substantial costs and personal morbidity. Fractures cause pain and disability. Hip fractures are associated with a 5–20 per cent excess mortality within the first year, and they often cause lasting functional disability. Each year 250,000 patients are admitted to hospital in the United States with hip fracture, incurring costs conservatively estimated to be at least $6 billion a year. In some countries more hospital beds are occupied by patients with hip fractures than by patients with acute myocardial infarction. Research programmes on osteoporosis have now given us several possible approaches for both preventing postmenopausal bone loss and treating established osteoporosis. Since therapeutic attitudes vary widely throughout the world, this conference was devoted to prophylaxis and treatment of osteoporosis.

1 Prevention of bone loss by oestrogen/progestogen administration

Oestrogen therapy prevents bone loss in postmenopausal women. It is currently the only well established prophylactic measure that reduces the frequency of osteoporotic fractures. Oestrogens have other effects, both favourable and unfavourable. Treatment with oestrogens alters fats in a way that may favour prevention of atheroma. There is evidence that oestrogen therapy may reduce the incidence of cardiovascular disease. The most important potential effect of oestrogen may be to reduce the risk of coronary heart disease. Oestrogens reduce or eliminate menopausal symptoms such as hot flushes, episodic sweating, vaginal dryness, and urethral irritation.

It is well documented that the risk of uterine cancer increases during and after oestrogen therapy. The risk is related to the dose and the duration of treatment. Well designed epidemiological studies do not suggest an overall increase in the risk of breast cancer in postmenopausal women treated with oestrogen. Cyclical or continuous administration of sufficient amounts of progestogens controls vaginal bleeding, eliminates endometrial hyperplasia, and reduces the risk of endometrial cancer.

The available evidence indicates that postmenopausal women identified to be at risk of developing osteoporosis should receive oestrogen therapy, provided there are no contra-indications and careful followup is ensured. Pending further information, the concomitant use of cyclic or continuous progestogen therapy is recommended in a woman with an intact uterus, in order to control bleeding and to reduce the risk of endometrial cancer. Treatment should be started as soon after menopause as possible. The benefits of oestrogen therapy are greatest just after the menopause, and intervention 15 to 20 years later is not desirable. The appropriate duration of oestrogen treatment is unknown, but at least 10 years seems to be reasonable.

2 Calcium in the prevention of osteoporosis

Nutritional intake of elemental calcium is an absolute requirement for bone health, facilitating growth and consolidation, and reducing bone loss after skeletal maturity has been reached. Estimates of calcium requirements for achieving and maintaining bone mass have yielded conflicting results, but a daily nutritional requirement of 800mg is recommended for European women, with higher requirements likely in childhood, in adolescence, and during pregnancy and lactation.

Calcium and peak bone mass

The amount of bone achieved at maturity (peak bone mass) is determined by genetic and environmental factors. Evidence from several studies suggests that dietary calcium intake in childhood and adolescence may be a determinant of peak bone mass. It has yet to be shown that decreasing the intake of calcium in childhood and adolescence results in a decreased peak bone mass. It seems likely that an adequate calcium intake is necessary for the maintenance of peak bone mass, but it is not known whether a high calcium intake at this stage of adulthood can contribute to improvements in bone mass.

Calcium and bone loss

Calcium in the diet does not substitute for oestrogen in preventing premenopausal bone loss. Some randomized clinical trials in postmenopausal women have found that a high calcium intake reduces the rate of loss of cortical but not of trabecular bone. Its efficacy in preventing fracture is uncertain. In postmenopausal women with established osteoporosis and women recognized to be at high risk of fracture it seems prudent for physicians to recommend a calcium intake of about 1500mg a day.

3 Exercise

Acute immobilization leads to loss of bone, which can be at least partially reversed by resumption of weight bearing activity. There is sufficient evidence to support the view that women should be encouraged to keep up a reasonable level of physical activity. Excessive exercise can lead to amenorrhoea and loss of bone. This very fact calls attention to the overriding importance of oestrogen in maintaining bone mass. The part played by exercise in reducing bone loss in the elderly is uncertain. It can be recommended in moderation, at least with the aim of improving patients' agility and reducing the likelihood of falls leading to fracture.

4 Fluoride

Fluoride may be used to increase trabecular bone mass in patients with severe vertebral osteoporosis. It is the only agent that has been shown to have a sustained effect on the formation of trabecular bone both at appendicular and at axial sites. Fluoride should not be used in patients with predominantly cortical osteoporosis; nor does it have a place in the prophylaxis of bone loss.

Fluoride increases trabecular bone mass, but some studies have shown directly or indirectly a decrease in cortical bone mass. Recent anecdotal reports of hip fracture associated with fluoride treatment

have emphasized this concern. It is prudent to reserve the
fluoride for patients with severe vertebral osteoporosis. Whether t
ment with fluoride reduces the rate of vertebral fracture is not known.

Up to one third of patients do not respond to treatment with fluoride
for reasons that are not yet known. In addition there are reversible effects — including gastrointestinal intolerance, bone pain, and arthritis
— in a substantial minority of patients sufficient to necessitate stopping
treatment.

There are some uncertainties about the optimum dose and duration
of treatment. A view consistent with general experience suggests that
doses of elemental fluoride of approximately 20mg daily are effective
but may cause defective mineralization, which can be overcome by
the concurrent administration of calcium supplements with or without
vitamin D. The optimum dose and duration of fluoride treatment are
not known, but the duration should probably not exceed five years.

5 Vitamin D and its metabolites and analogues

Vitamin D, its analogues, and its metabolities have been assessed both
in the prevention of bone loss and in the treatment of established
osteoporosis. In the early postmenopausal period there is no evidence
that vitamin D, its analogues, or its metabolites decrease the rate of
bone loss or fracture. There are conflicting and unresolved observations concerning the effect of vitamin D metabolites in decreasing the
rate of bone loss in patients with established osteoporosis.

6 Anabolic steroids

Currently available anabolic steroids do not have a place in preventing
osteoporosis in women. Although they can increase bone mass in
women with established osteoporosis, perhaps by increasing bone formation, serious concerns remain about their safety. Potential side effects that limit their use include virilization, a potentially atherogenic
effect on plasma lipoproteins, liver dysfunction, and the possible
development of liver tumours. Male sex hormones do have a place in
preventing and treating osteoporosis in some males.

7 Calcitonin

The use of calcitonin may be considered in the primary prevention of
osteoporosis in women at high risk who are not candidates for
oestrogen therapy, although its efficacy in this context has not been extensively studied. Calcitonin may be effective in reducing subsequent
bone loss in patients with established osteoporosis, but there is not
substantive evidence that it reduced the frequency of fractures.

Index

(Entries in **bold** type refer to tables)